Transactions on Intelligent Welding Manufacturing

More information about this series at http://www.springer.com/series/15698

Shanben Chen · Yuming Zhang · Zhili Feng
Editors

Transactions on Intelligent Welding Manufacturing

Volume I No. 3 2017

Springer

Editors
Shanben Chen
Shanghai Jiao Tong University
Shanghai
China

Zhili Feng
Oak Ridge National Laboratory
Oak Ridge, TN
USA

Yuming Zhang
Department of Electrical and Computer
  Engineering
University of Kentucky
Lexington, KY
USA

ISSN 2520-8519   ISSN 2520-8527   (electronic)
Transactions on Intelligent Welding Manufacturing
ISBN 978-981-10-8329-7   ISBN 978-981-10-8330-3   (eBook)
https://doi.org/10.1007/978-981-10-8330-3

Library of Congress Control Number: 2018933001

Printed on acid-free paper

This Springer imprint is published by the registered company Springer Nature Singapore Pte Ltd.
part of Springer Nature
The registered company address is: 152 Beach Road, #21-01/04 Gateway East, Singapore 189721, Singapore

Editorials

With the recent breakthroughs in deep machine learning and artificial intelligence and heightened demands on advanced manufacturing around the world, intelligentized welding and allied manufacturing processes, including robotic welding, will be a critical technology area and have been the recent development trends. This third issue of the Transactions on Intelligentized Welding Manufacturing (TIWM) presents recent developments in methods and technologies that lead to innovative intelligent welding applications. It includes two feature articles and eight papers selected from the 2017 International Workshop on Intelligentized Welding Manufacturing (IWIWM2017) contributing to intelligent welding manufacturing through understanding, sensing, and control of welding manufacturing processes.

The first featured article in this issue, "Thermal-Metallurgical-Mechanical Analysis of Weldment Based on the CFD Simulation", is by Jason Cheon and Suck-Joo Na from the Department of Mechanical Engineering, Korea Advanced Institute of Science and Technology (KAIST), Republic of Korea. Professor Na has been with the KAIST since 1983, where he is currently a professor emeritus. As a leading expert of CFD simulations of arc welding and laser materials processing, Dr. Na is Fellow of the American Welding Society and a member in the Korean Academy of and Technology. He is the recipient of the Humboldt Research Award, for his lifetime achievements. This feature article presents a new method of numerical thermal-metallurgical-mechanical analysis toward better understanding of welding processes. It simulates the temperature and phase fraction history using a CFD-FEM framework. Residual stress distribution of carbon steel weldment can be predicted as well.

The second featured article "Liquid Metal Embrittlement of Galvanized Steels During Industrial Processing: A Review" is a contribution from Shanghai Jiao Tong University. Liquid metal embrittlement (LME) is an important subject for automotive industry as higher strength steels are being introduced to reduce the weight and improve the crashworthiness of auto body structures. It focuses on the solid steel–liquid zinc material combination, which is highly relevant to automotive industry. The occurrence of LME in the process of industrial production such as hot-dip galvanizing, hot stamping, and welding are summarized.

The first selected paper of research articles, "Data-Driven Welding Expert System Structure Based on Internet of Things", by Chao Chen, Na Lv and Shanben Chen from Shanghai Jiao Tong University. The paper introduces a structure of data-driven welding expert system based on IOT and demonstrates its function. The data-driven welding expert system can learn and summarize the expert knowledge from these raw welding data without interacting with the welding expert.

The second paper, "Point Cloud Based Three-Dimensional Reconstruction and Identification of Initial Welding Position", is contributed by researchers from Shanghai Jiao Tong University. The authors propose a point cloud-based approach to recognize working environment and locate welding initial position using laser stripe sensor. Identification of workpiece is implemented by segmenting workpieces from the point cloud data in the image. Before segmentation, KD-tree-based background model is constructed to filter out background points; then RANSAC fitting procedure rejects outliers and fits the correct workpiece plane model; and the welding initial position can be found along the weld seam which is the intersection of fitted planes.

The third paper titled "A Robot Self-learning Grasping Control Method Based on Gaussian Process and Bayesian Algorithm", is contributed by a joint research team from Beihang University, Chongqing University of Science and Technology, and Wuhan University of Science and Technology. The paper presents a trained Gaussian process model combined with Bayesian algorithm. This method avoids the complex visual calibration process and inverse kinematics which are typically only applicable to a small group of samples. Besides, when the environment of grasping changes, the previous learning experience can be used to perform self-learning, and adapt to the grasping task in new environment, which reduces the workload of operators.

The fourth paper, "Real-Time Implementation of a Joint Tracking System in Robotic Laser Welding Based on Optical Camera", is from Shanghai University of Engineering Science. The authors introduce laser welding, describes composition of a real-time joint tracking system which mainly includes image acquisition part, image process and analysis part, and motion control part, reviews relevant investigations of joint tracking system and algorithm.

The fifth paper is "Effects of Process Parameters on the Weld Quality During Double-Pulsed Gas Metal Arc Welding of 2205 Duplex Stainless Steel", the waveform charts of the current and voltage, the U-I diagram, the energy input, the dynamic resistance, and the other real-time signals were analyzed. In addition, the mechanical tensile and metallographic tests were executed. The results demonstrated that the welding speed had the highest impact on the welding quality among all three factors, which was followed by the number of weak pulses and the number of strong pulses had the least impact.

The sixth paper, "Grain Boundary Feature and Its Effect on Mechanical Property of Ni 690 Alloy Layer Produced by GTAW", is a contribution from Shanghai Jiao Tong University. Nickle-based Alloy 690 surfacing layers were fabricated by gas tungsten arc welding (GTAW) with two different heat inputs, namely, large heat input (LHI) and small heat input (SHI). It is found that the ultimate tensile strength (UTS) of the LHI samples was higher than that of the SHI samples after reheat thermal cycles, regardless of the reheating temperature. The EBSD result shows that the proportion of high angle grain boundaries (GBs, >15°) in the LHI sample was obviously higher than that in the SHI sample.

The seventh paper is titled as "Migration Behavior of Tungsten Carbide in the Dissimilar Joints of WC-TiC-Ni/304 Stainless Steel Using Robotic MIG Welding". WC migration and the formation of gradient layers are discussed by presenting a self-sealing model. The results show that WC migration not only happens in the fusion zone, but also in the HAZ, especially near the top surface, which led to gradient layer, η phase, and tungsten dissolution–re-precipitation on the surface of WC. The results also indicate that the fusion zone has the ability to cure the cracks itself during the robotic MIG welding.

A short paper, "Study on Properties of LBW Joint of AISI 304 Pipes Used in Nuclear Power Plant", discusses laser beam welding of a girth weld of AISI 304 stainless steel pipes. Microhardness test was carried out to determine the mechanical properties of the welded joint. A slow strain rate test was performed to investigate the susceptibility to stress corrosion cracking (SCC) of the welded joint in the simulated pressurized water reactor environment. The results show that the values of microhardness in the weld and HAZ are close to those in the parent material.

Welding is a critical manufacturing technology in the modern industry. Sustained research is essential to establish the foundation for intelligent welding processes and systems for quality and productivity.

Zhili Feng Ph.D.
TIWM Editor-in-Chief
Oak Ridge National Laboratory, USA
fengz@ornl.gov

Contents

Part III   Short Papers and Technical Notes

Part I
Feature Articles

Thermal-Metallurgical-Mechanical Analysis of Weldment Based on the CFD Simulation

Jason Cheon and Suck-Joo Na

Abstract A new method of numerical thermal-metallurgical-mechanical analysis was introduced in this paper. The CFD welding simulation is based on the mass and heat transfer analysis solving mass, momentum, and energy conservation equations along with the Volume of Fluid (VOF) method. The VOF method is employed to track the shape of the free surface. The arc and droplet heat source model with electromagnetic force and arc pressure model were used for the arc welding process. Next, the temperature history of CFD welding simulation was transferred to the FEM domain for thermal-metallurgical-mechanical analysis with CFD-FEM framework. The diffusion kinetics considered phase transformation model successfully predicted phase fraction and residual stress distribution of carbon steel weldment. By using the combination of suggested T-M-Me analysis method and CFD welding analysis, it is possible to reproduce a phenomenon closer to reality. Also, the recent CFD-based process analyses and results that can be extended to multi-physical analysis were briefly introduced. However, considerable assumptions and simplified models are different from real welding phenomena. To solve this gap and to use welding simulation as a prediction tool rather than a reproduction, many young researchers will need to challenge.

Keywords Computational fluid dynamics · Process model
CFD-FEM framework · Data transfer scheme · Phase transformation
Residual stress

1 Introduction

Numerically predicting a phase transformation and its mechanical effect on carbon steel, begins with a heat treatment process analysis [1]. By considering carbon steel as a solid solution in the diffusion process, the phase fraction will be affected by

J. Cheon · S.-J. Na (✉)
Department of Mechanical Engineering, KAIST, 291 Daehak-ro, Yuseong-gu,
Daejeon 34141, Republic of Korea
e-mail: sjoona@kaist.ac.kr

© Springer Nature Singapore Pte Ltd. 2018
S. Chen et al. (eds.), *Transactions on Intelligent Welding Manufacturing*,
Transactions on Intelligent Welding Manufacturing,
https://doi.org/10.1007/978-981-10-8330-3_1

3

chemical composition and temperature history during the cooling process. Consequently, the heat treatment process analysis is based on a time-temperature-transformation (TTT) diagram or continuous cooling transformation (CCT) diagram data for each type of carbon steel. Moreover, for precise estimation of the temperature history, the analysis also focuses on cooling models, including convection and radiation heat loss to the cooling medium. By adding a heating process model, including a heat source, and austenization phase prediction, phase transformation prediction analyses of welding processes have been conducted using the finite element method (FEM) based conductive heat transfer (CHT) analysis [2–6]. According to previous thermal metallurgical welding stress analyses, accounting for the phase transformation effect in the mechanical analysis is important, when considering a welded high carbon steel structure. The residual stress is highly affected by the phase transformation induced strain component, and the phase transformation is highly affected by thermal history.

As a numerical thermal analysis method, FEM-based CHT analysis has been widely used with Gaussian surface flux distribution or double ellipsoidal power density distribution. The double ellipsoidal power density distribution was suggested by Goldak et al. [7] in 1984 as a heat source to reproduce the penetrating action of the arc and droplet, which transports heat well below the surface. Because of their simplicity and accessibility, Gaussian surface flux distribution (or surface heat source, SHS) and double ellipsoidal power density distribution (or volumetric heat source, VHS) are today the most widely employed for analysis of the arc welding process.

However, the FEM-based CHT analysis with VHS has physical limitations in the GMAW process. First of all, artificial thermal properties for the liquid state are required, due to the absence of the advection effect. In the GMAW process, the mass flow of droplets and the arc force driven convective flow on the molten pool diminishes the fusion-zone temperature. The FEM-based CHT analysis achieves a diminished fusion-zone temperature by enhancing the diffusion effect with artificial thermal properties. Nonetheless, the FEM-based CHT analysis with VHS for the GMAW process has been employed until today based on its advantages, which include reasonable calculation speed, the simplicity of the heat source, and good accessibility to the thermo-mechanical analysis.

Meanwhile, computational fluid dynamics (CFD) based mass and heat transfer (MHT) analyses have also been employed for the numerical thermal analysis of welding processes since the early 2000s. The main merit of CFD-based MHT analyses is the ability to consider the advection effect on the molten pool. Because of this advantage, it has been possible to consider the momentum of the molten pool flow, process dependent bead shape, and reasonable molten pool temperature without the use of artificial thermal properties of the liquid state.

Following the development of the image recording technique, the parameters of SHS have been extricated from uncertainty using the measuring arc intensity distribution scheme with high-speed camera image analysis, using the Abel inversion method [8] and Fowler-Milne method [9]. Also, droplet mass and the heat source's parameters (size, velocity, and initial position) were changed from a rough assumption to measurement based information. However, complex force and pressure models are still required to consider momentum balance with mass balance in the CFD-based MHT analysis. This complexity degrades accessibility and computational calculation speed. Also, the CFD-based MHT analysis result makes it impossible to extend the thermo-mechanical analysis without a special data transfer scheme, even with a reasonable result.

Phase transformation data (TTT and CCT) are built from the constant temperature condition or constant cooling rate condition. The welding process does not occur at a constant temperature and or constant cooling rate, but previous researchers [3–5, 10] have made prediction of phase transformations with additional assumptions and hypothesis. The main target has been the prediction of weldment deformation, but the phase fraction of the weldment has not been validated yet.

The representative numerical stress analysis method used for considering phase transformations during thermal processes is the quenching process stress analysis [11–13]. The phase transformation induced strain values from austenite to bainite or martensite have been experimentally of analytically determined. On the other hand, for the welding process, the phase transformation stress analysis involves both a heating and cooling process. In previous numerical phase transformation considered mechanical analyses, the stress concentration is determined by the induced strain of the phase transformation, and the phase dependent mechanical properties [2–5, 14–16]. Dilatometry data for a constant heating and cooling rate condition has been used to define the dilatometric behavior produced by transformations during the gas tungsten arc (GTA) spot welding process [4], and the hard-face-welding process [14]. Cho and Kim [2] suggested that volumetric strain was induced by a phased fraction based transformation. In the heating process, the ferrite to austenite transformation from AC1 to AC3 produces a phase transformation induced volumetric (TRIV) strain, instead of a thermal expansion strain. In the cooling process, the austenite to martensite transformation was considered with TRIV strain. Also, the phase fraction of martensite was estimated using the welding time constant, Δt_{8-5}. Deng and Murakawa [10] compared the effects of TRIV and phase transformation dependent yield strength on the welding residual strain for the GMAW process. They evolved a transformation model with considering the austenite to bainite transformation [15]. Finally, the transformation induced plastic (TRIP) strain effect [17, 18] was reproduced in the recent literature [5, 16].

In their previous research on in welding procedures, a phase transformation based mechanical analysis was applied to real welding phenomena. However, their model has the potential for further development. If a more reliable metallurgical prediction model can be applied to the mechanical model, the results of a thermal metallurgical mechanical analysis of the weldment would contribute to a better understanding of its natural behavior. Also, a more realistic thermal analysis might lead to more realistic results for the whole procedure.

For extension from the CFD-based MHT analysis result to metallurgical and mechanical analysis of weldment, CFD-FEM framework was presented [19]. By transferring the temperature history from CFD to the FEM domain, the thermal history of a CFD-based MHT analysis can now be used for multi-physics analysis.

This paper presents a brief summary of recent numerical approaches for thermal-metallurgical-mechanical (T-M-Me) analysis of GMAW weldment based on the CFD-FEM framework [19–21]. Also, the CFD based analysis results of various fields that can be extended to multi-physics analysis are introduced.

2 Experiment and Measurement

Table 1 gives the welding conditions used to produce a bead on plate (BOP) welding of $160 \times 89 \times 6$ mm^3 (thickness) AH36 steel plate using the GMAW process. Figure 1a is the schematic of the welding process. Figure 1b

Table 1 Experiment conditions

Feed rate m/min	Current (A)	Voltage (V)	CTWD (mm)	Welding speed (mm/s)	Electrode diameter (mm)	Shielding gas 80% Ar-20%CO$_2$ (L/min)
7.5	237.7	26.9	28	10	1.2	20

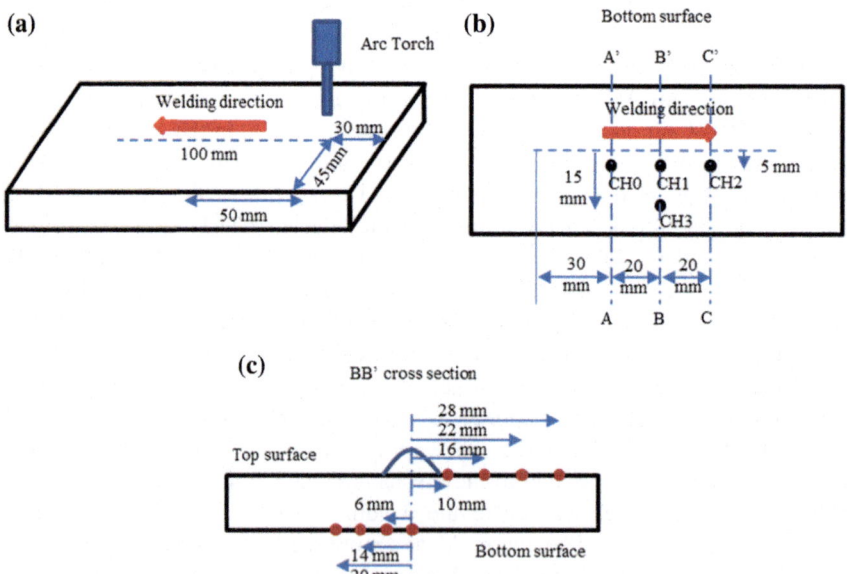

Fig. 1 **a** Schematic of experiment setup; **b** cutting plane and thermocouple position; and **c** positions used to measure directional stress on the BB′ cross section

explains the thermal cycles and the locations used for measuring micro hardness. The instantaneous welding current and voltage data were measured. The welding arc images and droplet transfer were captured using a high-speed camera. The effective radii of arc were extracted from the arc plasma images using the Abel inversion method [35] and the Fowler-Milne method [36]. The droplet information was also extracted by image pixel counting. The weldment cross-sections were cut along the AA′, BB′, and CC′ lines. Next, the cross-section was etched with 2% Nital solution to see the fusion zone (FZ) and the heat affected zone (HAZ). Further, the hardness distribution was measured. Additionally, the BB′ cross-section was etched with Vilella's etchant to distinguish ferrite regions [22]. Figure 1c explains the locations used for measuring directional stress on the top and bottom surfaces.

3 Thermal Analysis

A set of the governing equation were solved with process models for the CFD-based MHT analysis in an incompressible laminar flow condition with Newtonian viscosity model, and calculated in CFD commercial software, Flow3D [23]. The governing equation included the continuity, momentum, energy, and the volume of fluid equation. The energy equation is

$$\frac{\partial h}{\partial t} + \bar{V} \cdot \nabla h = \frac{1}{\rho} \nabla \cdot (k \nabla T) + h_s \tag{1}$$

The advection term (2nd term of left hand side) restrains the increase of the unsteady term (1st term of left hand side) by the diffusivity term and source term (right hand side terms). Once the temperature becomes lower than T_{sol} (or does not exceed it), the advection term returns to zero and it behaves the same as conductive heat transfer. The process models were the arc heat source as a form of double ellipsoidal surface heat source, droplet mass and heat source, arc pressure model, electromagnetic force, drag force, and heat loss model. The detailed procedures and the model descriptions in this chapter have already been explained elsewhere [8, 9, 24]. To ensure agreement of the coordinates and directions of the process model, a fully modeled domain was employed, as shown in Fig. 2.

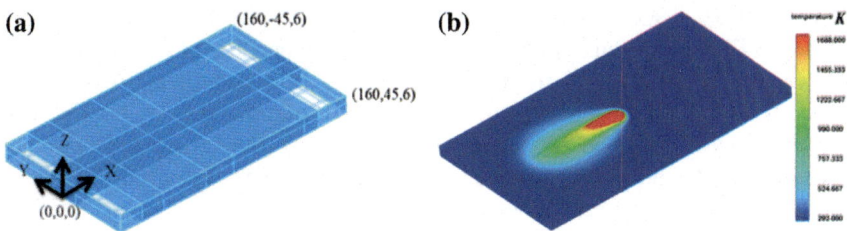

Fig. 2 **a** Solution domain of CFD analysis with coordinate system; **b** temperature distribution at $t = 5$ s

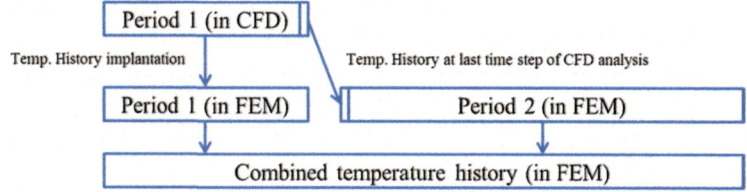

Fig. 3 Schematic description of CFD and FEM combined thermal analysis of GMAW process

	CFD	Method	FEM	
	Control Volume (CV)	Volume	Element	
	Control Volume Center (CVC) Center of CV	Data Point	Node Vertex of Element	

Fig. 4 Schematic descriptions of volume and data points for CFD and FEM configuration

The temperature history was incorporated from the results of two thermal analysis methods for the better accuracy and shortened computational time (Fig. 3). CFD mass and heat flow analysis results were adopted from welding start to solidification (period 1) and FEM conductive thermal analysis results were employed for the cooling process (period 2). As the initial condition for period 2, the temperature distribution of the end step of the period 1 CFD result was applied to the whole solution domain. The same thermal properties were used for continuity of temperature history.

As shown in Fig. 4, the word "node" is a vertex of a polygonised "element" in FEM [25]. The physical and mathematical information were calculated and stored in the node, and the same was also propagated from node to node. On the other hand, the basic unit of configuration in CFD is "control volume" [26]. The calculation and the storing of physical and mathematical data were processed in the Control Volume Center (CVC). For the same size solution domain, the position of the node and CVC are not the same in any density of element or CV. Because of this difference between the two numerical methods, proper data transfer is not possible without a proper data implantation scheme.

To get rid of the mismatched spatial location problem between the node and CVC, a Binary Search Algorithm was employed [27]. In the 3D analysis, the number of CVC's surrounding node is eight. The temperature of a node is calculated by linear interpolation from the eight CVCs (Fig. 5a). In the temperature history of the CFD simulation, spatial temperature distribution data was stored with each discretized time step. Assuming that the temperature fluctuation was not severe within the time step, a linear interpolation scheme was used. Figure 5b represents a schematic of the temporal interpolation.

The Abaqus subroutine was used to achieve temperature history implantation from CFD to FEM in this study. Additionally, a coordinate rotation matrix was used

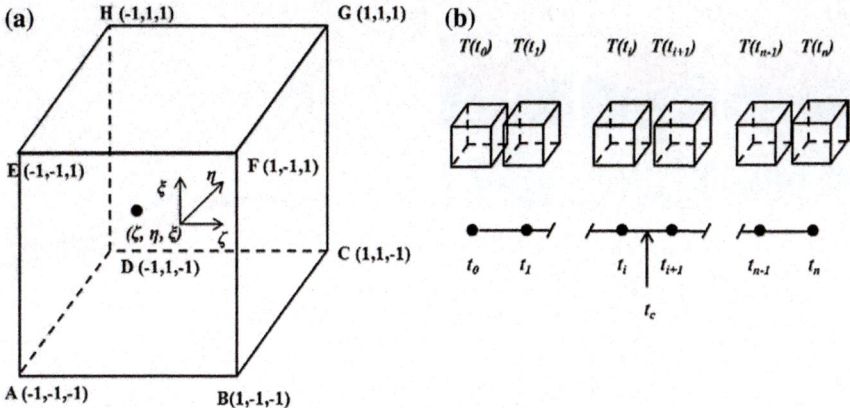

Fig. 5 **a** Schematic diagram of node and surrounding CVCs; **b** schematic diagram of temporal interpolation

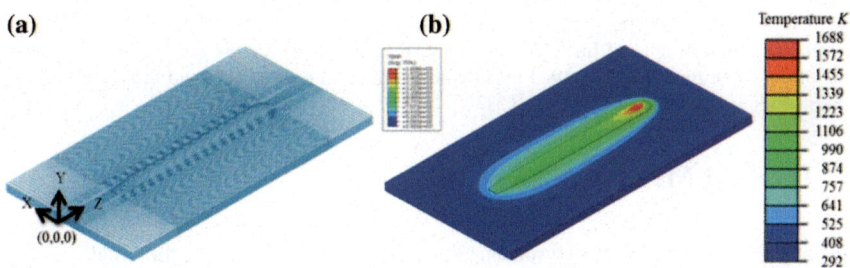

Fig. 6 **a** Solution domain of FEM analysis with coordinate system; **b** temperature distribution at $t = 10.9$ s (start of period 2)

to take care of the change in domain axis between CFD and FEM (Figs. 2a and 6a). The CFD temperature history was implanted into the FEM solution domain for period 1, and the last time step temperature distribution was set as an initial condition for the FEM conduction heat transfer analysis of period 2, the cooling process.

Figures 7a–c depict the weld pool profiles of the simulation results. The corresponding measured weld micrographs at sections AA′, BB′ and CC′ (as marked in Fig. 1b) are also shown. The red colored area represents the fusion zone of the weldment (peak temperature $\geq T$s). Figure 7d presents the variation between the experimental and the calculated weld bead dimensions at the above mentioned three sections. It can be observed that the fusion zone of the weldment has a fingertip shape with slight under-cuts at the end of each bead. As the welding progressed, the penetration depth was increased while the reinforcement height was decreased. The calculated fusion zone profile and its dimensions were in fair agreement with the corresponding experimental results. Figure 7e shows fair good agreement of analysis on temperature history at the four positions marked on the bottom surface of the base plate (Fig. 1b).

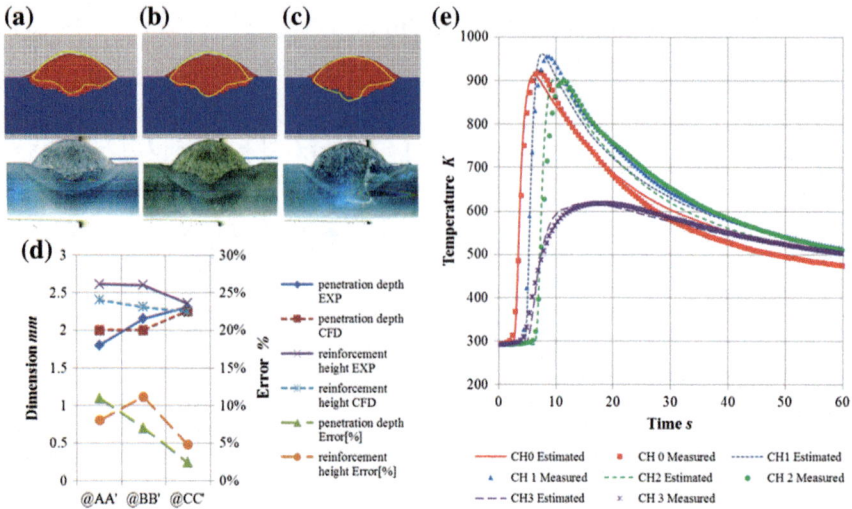

Fig. 7 Simulation and experimental results for the cross-section of **a** AA′; **b** BB′; **c** CC′ (the red colored area is the simulated fusion zone, and solid yellow lines are the boundary of the fusion zone of the experiment); **d** quantified result comparison, and **e** calculated and the corresponding measured temperature history of CH0–CH3

4 Thermal-Metallurgical Analysis

The phase transformation estimation algorithm [19] is based on the fraction conservation relation as below

$$F_{total} = F_f + F_a + F_b + F_m + F_l \tag{2}$$

F_{total} and F_i represent the total fraction sum and the fraction of the i phase. The subscripts f, a, b, m, and l refer to the ferrite, austenite, bainite, martensite and liquid phases. As an initial condition, 100% ferrite fraction was assumed. The existence of the retained austenite and acicular ferrite were ignored to avoid the complexites in the kinematic behavior, as shown in the CCT diagram of AH36 steel [6] (Fig. 8a).

The austenite fraction in the heating process was estimated using linear interpolation between AC1 and AC3. AC1 and AC3 were calculated as the fitted power functions of the instant maximum heating rate (HR) with experimental data [28, 29]. The CCT diagram information including the starting temperature, finishing temperature and the maximum phase fraction at a certain cooling rate for each phase transformation were fitted as the price-wised continuous polynomial functions of the instant cooling rate (CR). Based on the fitted CR dependent transformation information in the cooling process (Fig. 8b), the austenite to ferrite and austenite to bainite fractions were calculated using linear interpolation between transformation starting and finishing temperature. The martensite transformation fraction was calculated by the Koistinen-Marburger (K-M) equation. Also, the hardness

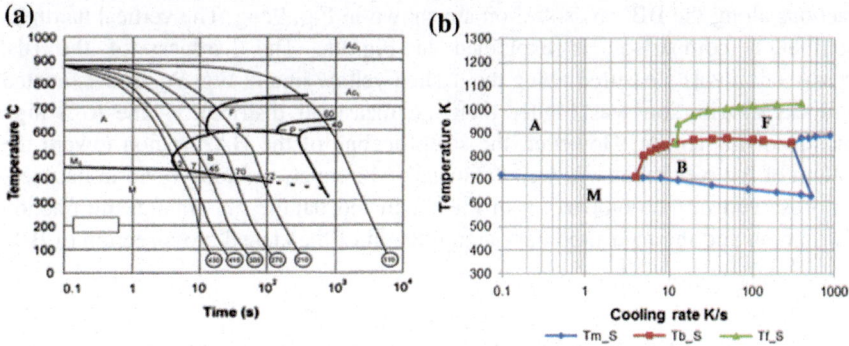

Fig. 8 **a** CCT diagram of AH36 steel and **b** starting temperature versus cooling rate plot for each phase

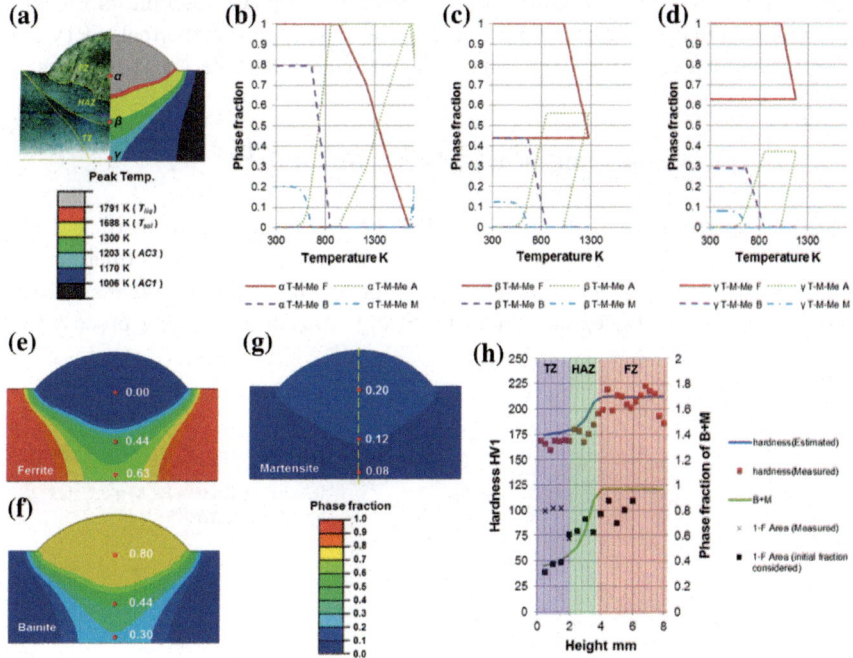

Fig. 9 Phase fraction versus temperature of **b** point α; **c** point β and **d** point γ of **a**, **e–g** are the resultant phase fraction distributions on the BB cross section, **h** is a vertical distribution comparison of the estimated versus measured hardness, and the bainite + martensite fraction along the dash-dot line in **g**

estimation was performed using the fitted polynomial function of the martensite fraction based on the CCT diagram hardness information.

The results of the metallurgical analysis is shown in Fig. 9. The phase fraction history of the α, β and γ points (Fig. 9a) are shown in Fig. 9b–d. The final phase fractions along the BB' cross-section are shown in Fig. 9e–g. The vertical hardness distribution comparison is explained in Fig. 9h. The hardness of the BB' cross-section was measured along the dashed yellow line in Fig. 9g. The estimated hardness distribution was higher than the measured distribution due to a high fraction of martensite. However, the steep decline of the HAZ region toward the bottom of the weldment was well estimated. Moreover, a similar trend in hardness was observed in the comparison of the calculated bainite plus martensite fraction distribution and the measured dark area fraction of the etched cross-section of BB'.

5 Thermal-Metallurgical-Mechanical Analysis

The 3-dimensional T-M-Me analysis [21] was conducted using a linear kinematic hardening model with the von Mises yield criteria. The phase fraction information was employed in each subroutine to calculate the elastic, plastic, thermal, TRIV and TRIP strains at every time step. The strain increment balance can be written as below

$$\Delta \varepsilon_{ij} = \Delta \varepsilon_{ij}^{e} + \Delta \varepsilon_{ij}^{p} + \Delta \varepsilon_{ij}^{TH} + \Delta \varepsilon_{ij}^{\Delta V} + \Delta \varepsilon_{ij}^{TRIP} \tag{3}$$

where the superscripts e, p, TH, ΔV and $TRIP$ are the elastic, plastic, thermal, TRIV and TRIP strains. The general behavior of thermal-metallurgical strain (thermal and TRIV strain) with respect to temperature during the welding process can be described using the conceptual schematic in Fig. 10a. In the heating process (red

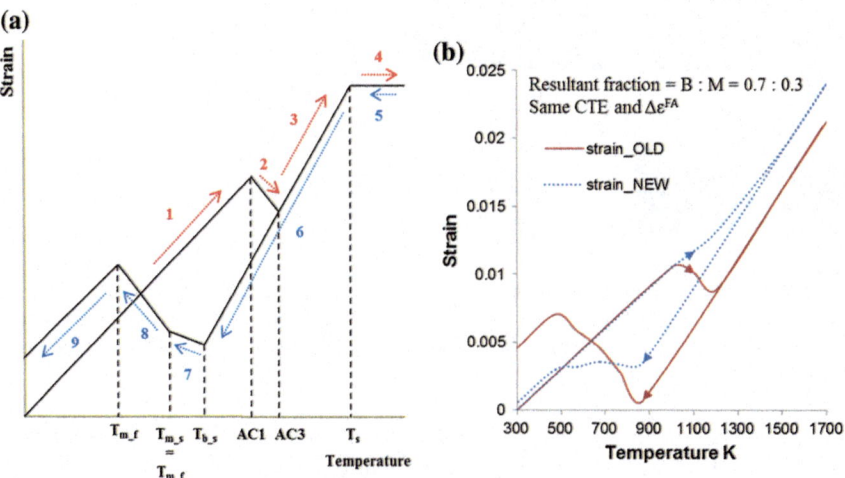

Fig. 10 **a** The schematic of thermal and TRIV strain behavior versus temperature and **b** strain versus temperature curve comparison between previous concept (strain_OLD, solid line) and this study's concept (strain_NEW, dotted line)

arrows), strain increases with increasing temperature, except for the ferrite to austenite transformation (2), while strain decreases with reducing temperature, except for the austenite to bainite transformation (7) and martensite transformation (9) in the cooling process (blue arrows). As noted in previous studies, if the transformation starting and finishing temperatures are fixed and the temperature range is not very wide, TRIV shrinkage (2) and expansion (7, 9) can be determined without considering the thermal expansion of each phase. However, we assumed that the *AC1* and *AC3* were dependent on *HR*, and that temperature range was wide. Therefore, the thermal strain has to be considered in the transformation temperature range. The new concept of overall thermal and TRIV strain behavior can be determined, as below, in relation to increasing temperature ΔT and the coefficient of thermal expansion (CTE) of i phase α_i.

$$\Delta\varepsilon^{TH} + \Delta\varepsilon^{\Delta V} = F_F\alpha_F\Delta T + F_A\alpha_A\Delta T + F_B\alpha_B\Delta T + F_M\alpha_M\Delta T$$
$$- \Delta\varepsilon^{FA}\Delta F_A + \Delta\varepsilon^{AF}\Delta F_F + \Delta\varepsilon^{AB}\Delta F_B + \Delta\varepsilon^{AM}\Delta F_M \tag{4}$$

The difference between the old concept from previous studies and the new concept of thermal and TRIV strain behaviors is demonstrated in Fig. 10b.

The resultant response (RES) strain and *YS* distribution on the top surface of the BB′ cross-section are shown in Fig. 11b, c comparing thermal-mechanical (T-Me) analysis results. The differences in the RES strain and *YS* were caused by the phase fraction distribution. Significant differences occurred in the FZ and its neighboring region. Eventually, the von Mises stress was concentrated in the same area on the top (Fig. 11d) and bottom (Fig. 11e) surfaces. A prominent nonlinear resultant

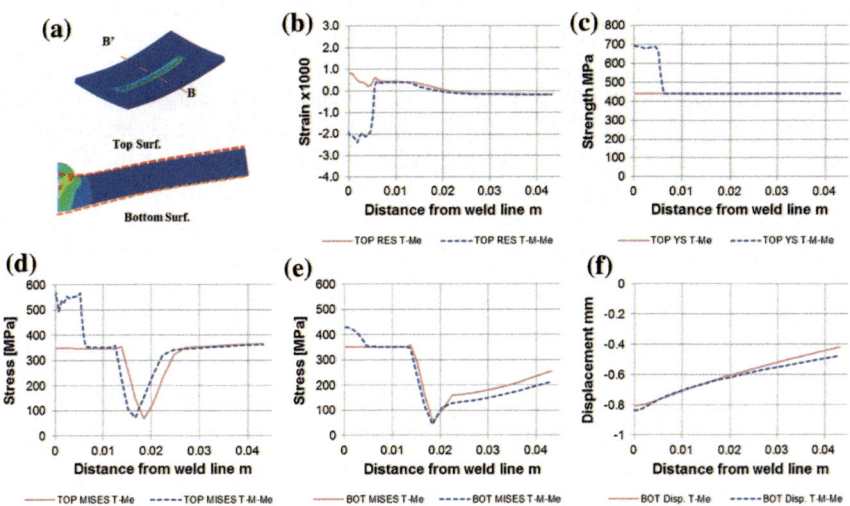

Fig. 11 **a** Is a schematic of the top and bottom surfaces of the BB′ cross-section; **b** and **c** are results for the **b** RES strain and **c** YS distribution on the top surface; **d** and **f** are the results of the von Mises stress distribution on **d** the top surface and **e** bottom surface; **f** is the vertical displacement (u_y) on the bottom surface

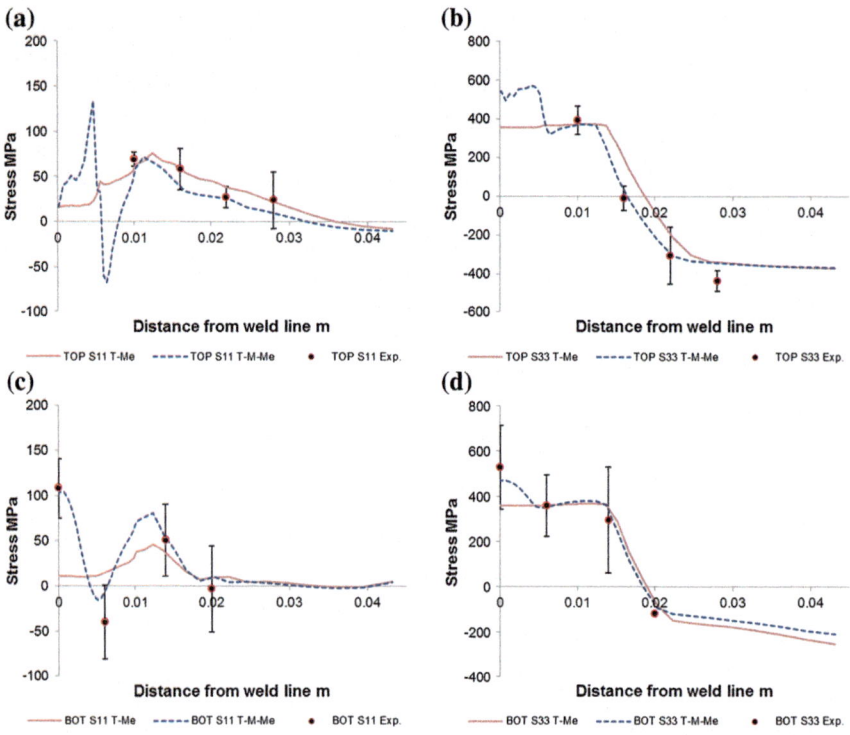

Fig. 12 Comparison of the calculated and measured values of directional stresses on the top surface (**a** transversal and **b** longitudinal) and the bottom surface (**c** transversal and **d** longitudinal)

vertical displacement of the bottom surface (Fig. 11f) was computed in the T-M-Me analysis.

The directional stress distributions on the top and bottom surfaces of the BB′ cross-section are plotted in Fig. 12 along with the values measured (averaged, black dots) at the points in Fig. 1c. The calculated stress distributions of the T-M-Me analysis were well followed measured values than the T-Me analysis. In particular, the nearness of the welding line's fluctuation of stress on the bottom surface confirms the validity of this study's T-M-Me analysis. The comparisons of Fig. 12 confirm the potentials of thermal-metallurgical-mechanical analysis based on the CFD simulation.

6   Extensity of Technique

The CFD-based process analysis can be extended to a multi-physics analysis by the previously described thermal-metallurgical-mechanical analysis technique. Of course, the calculation of the full domain is necessary, and the metallurgical and mechanical properties of the material must be accompanied. This chapter introduces recent CFD-based process analysis techniques and results that can be extended to multi-physical analysis.

6.1 Arc Welding in Various Conditions

For V-groove welding, the coordinate mapping technique for electromagnetic force model and elliptical symmetric arc model were suggested [30]. Also, asymmetric arc model was adopted for the second pass welding on V-groove weldment [31]. Meanwhile, the process models were formulated in a rotated coordinate system for horizontal fillet joints [32]. The numerical analysis results show high accuracy in various welding speed and wire feed rate conditions (Fig. 13).

Considering heat transfer from arc plasma, droplet and molten slag to material surface, submerged arc welding (SAW) process was simulated [33]. Also two wires [34] and three wire [35] tandem SAW process were well modeled employing arc interaction model. Moreover, the molten pool behavior by flux-wall guided metal transfer was revealed by the predefined solid cavity component [36] (Fig. 14).

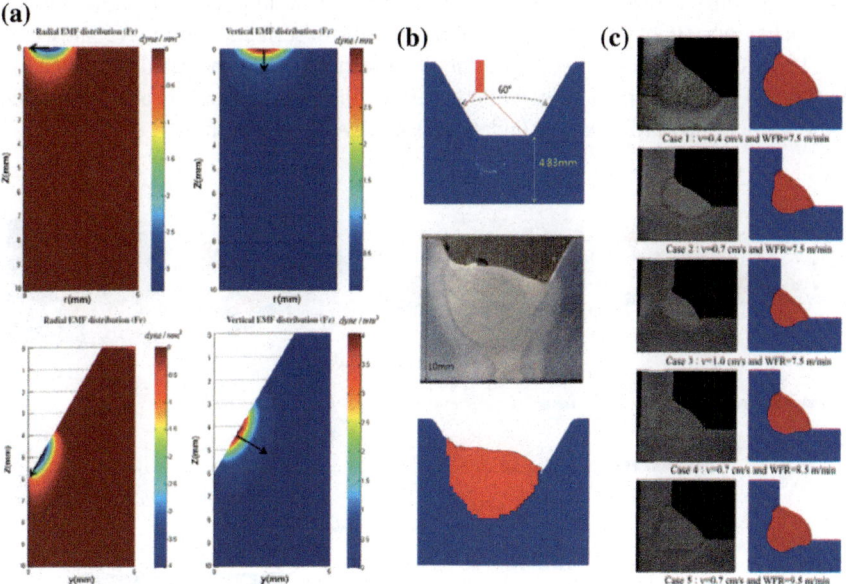

Fig. 13 **a** Comparisons of electromagnetic force of BOP and V-groove feature with mapping technique; **b** schematic and results of experiment and simulation on second pass welding on V-grove; and **c** fusion zone shape comparison of horizontal fillet welding in various process conditions

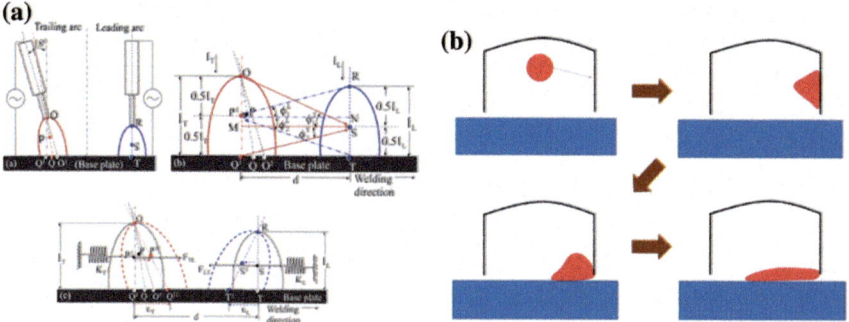

Fig. 14 Schematics of **a** arc interaction model for tandem process and **b** flux-wall guided metal transfer

6.2 Visualization Using Transient Simulation Result

Solving unsteady terms of governing equations, the results of CFD based simulation can be utilized for welding phenomena visualization. A numerical research was performed for circumferential welding process on the pipe V-groove in various welding positions by changing gravity direction and root gap [37]. The numerical models not only formed a stable weld bead but also simulated the dynamic molten pool behaviors such as overflow which was not described before. As shown in Fig. 15, without the root gap, it is difficult to form a fully penetrated weld bead in the flat and overhead positions, while humping and melt-through beads are formed in the vertical-upward position under the same welding condition.

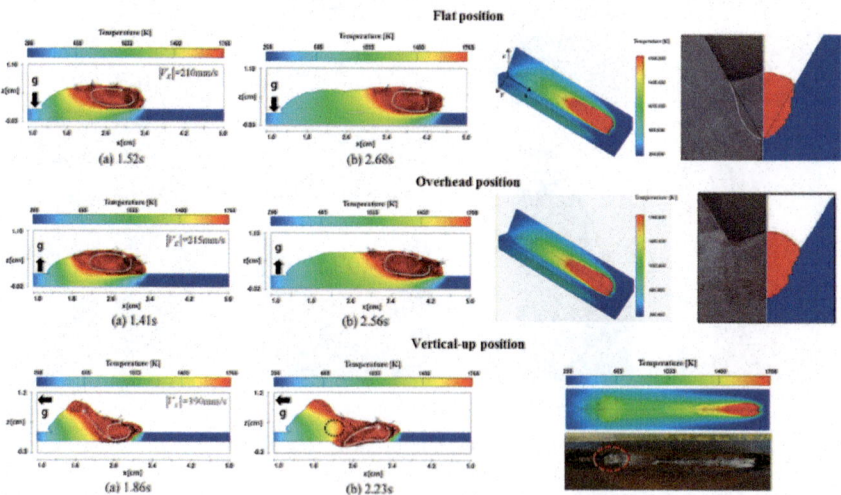

Fig. 15 Flow patterns and comparison simulated bead shape and experiment for non-root gap cases

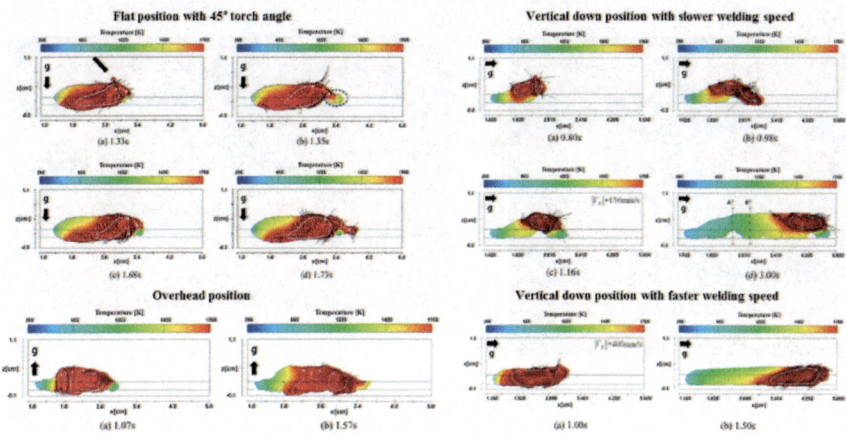

Fig. 16 Flow patterns for 1 mm root gap cases

With a 1-mm root gap as shown in Fig. 16, the molten pool overflow patterns can be described for various welding positions under the given welding conditions. The overflow patterns in some welding positions do not induce the weld defects, while a weld bead with incomplete penetration can be formed in the vertical downward position. Thus, it is necessary to avoid the overflow patterns in such a case by increasing the welding speed.

Meanwhile, a transient welding simulation result was used to visualize momentum flow in the produced fingertip shaped molten pool [38]. In the direct current process, current and voltage are maintained at a constant value. Consequently, only the arc pressure and EMF are used to maintain the molten pool flow. Besides that, the droplet impingement is used to agitate the molten pool flow with a given frequency, even in the DC process. Therefore, investigating the molten pool flow at one droplet intervals could be help to understand the formation of fingertip penetration. Molten pool speed information was utilized to track momentum flow in the weld pool using color maps and streamline plots in various planes (Figs. 17, 18 and 19).

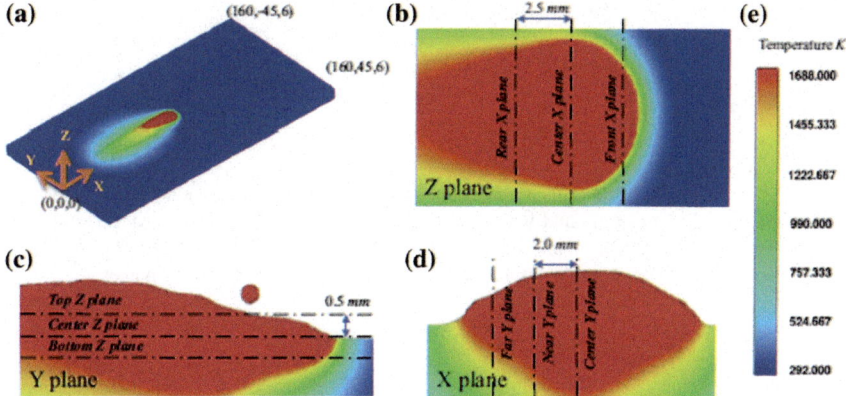

Fig. 17 Analyzed planes in momentum visualization study

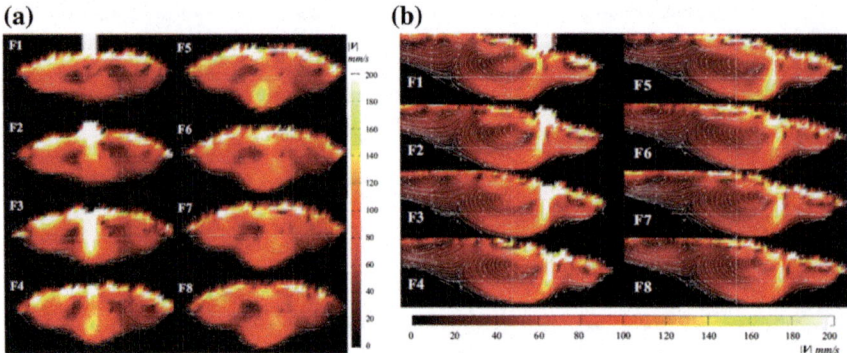

Fig. 18 Momentum flow history of **a** Center X plane and **b** Center Y plane

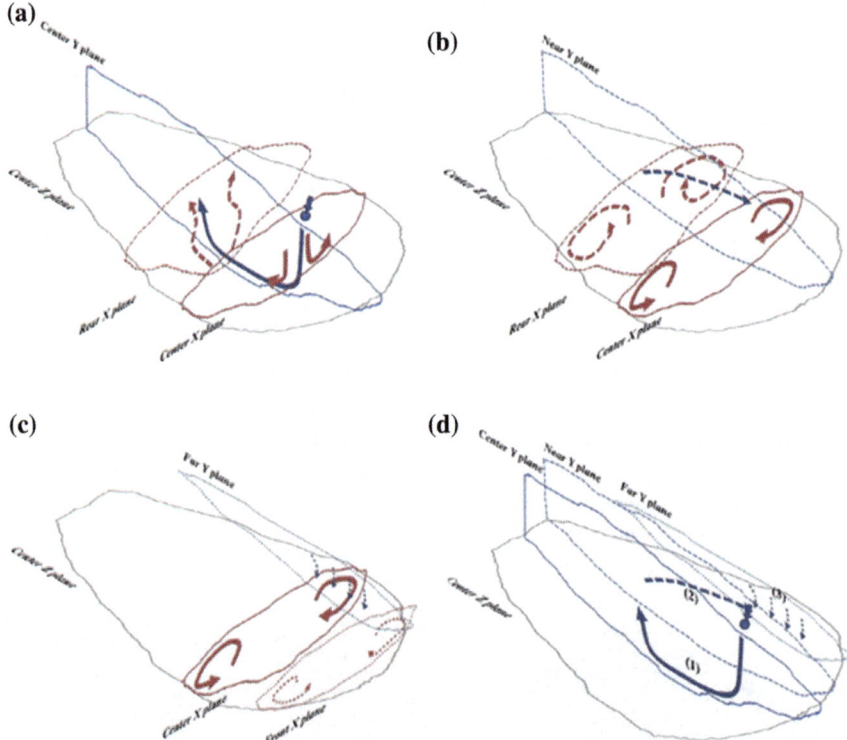

Fig. 19 The schematics of droplet impingement momentum flow on the molten pool, **a** the first effect of the droplet momentum and **b–c** secondary effect of droplet momentum (blue arrows are the strongest flow pattern in the Y-planes, and red arrows are the most active flow patterns in the X-plane); **d** sequential description of droplet impingement momentum flow on Y-planes

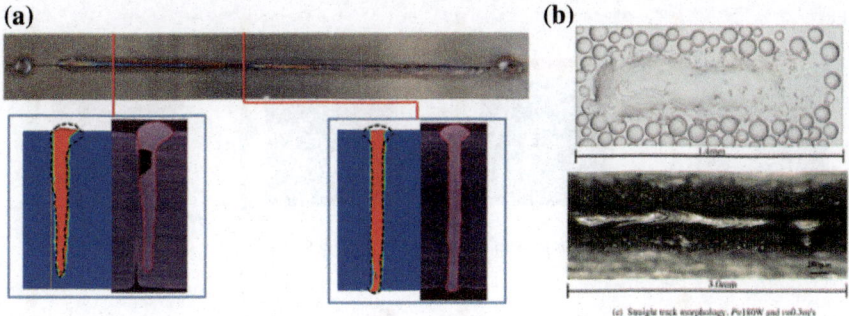

Fig. 20 Comparisons of experiment and simulation for **a** non-zero gap laser butt welding and **b** single layer morphology of selective laser melting process

As results, it was determined that the droplet impingement momentum first strikes the bottom of the molten pool and digs a deep fingertip penetration. Then, the droplet impingement momentum detours backward at a deep level and moves forward at a shallow level, and widens the molten pool width as shown in Fig. 20.

6.3   Laser Processes

The Fresnel's reflection model for multiple reflections of lay with recoil pressure, scattering, and absorption model is utilized for the laser welding process and selective laser melting process. The laser matter interaction techniques were utilized for laser ablation on semi transparent material [39], laser welding process [40–42], and selective laser melting process (SLM) [43].

6.4   Arc-Laser Hybrid Welding Process

The arc and laser welding processes are the most widely used. Great effort is required to understand the physical phenomena of arc and laser welding due to the complex behaviors which include liquid phase, solid phase and, gas phase. It is necessary to conduct a numerical simulation to understand the detailed procedures of welding. The research was presented the effects of the various potential force models on laser–arc hybrid welding [44] (Figs. 21 and 22).

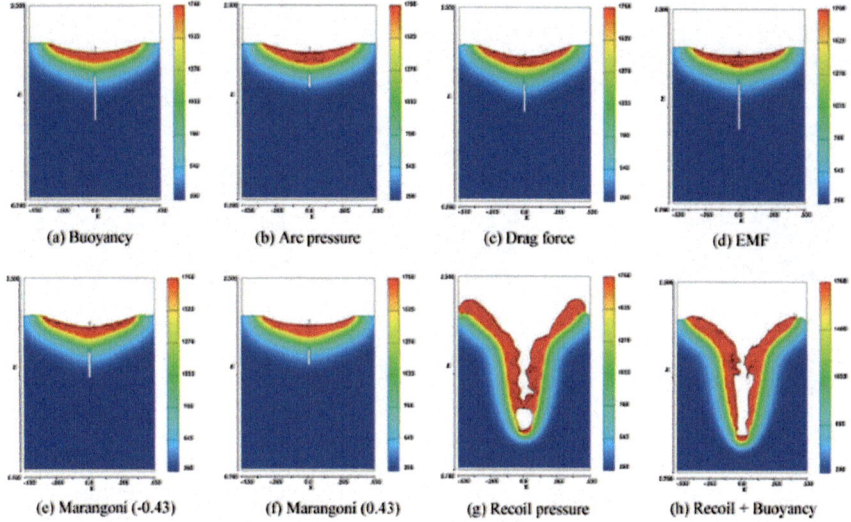

Fig. 21 Effects of potential force models on laser-arc hybrid welding 1

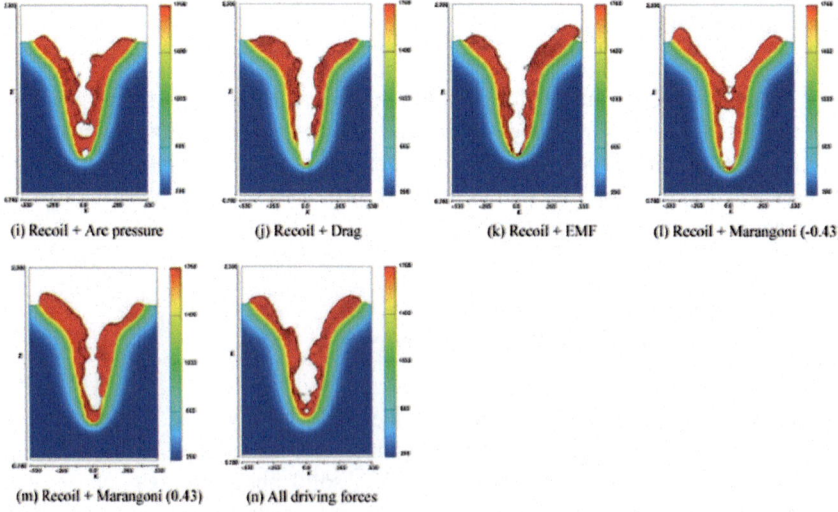

Fig. 22 Effects of potential force models on laser-arc hybrid welding 2

7 Conclusion

A new method of numerical T-M-Me analysis was introduced in this paper. The thermal-metallurgical analysis method was employed in the T-M-Me analysis using real-time temperature and phase fraction history based on the CFD-FEM framework. The suggested core elements of analysis model are summarized below.

- Measurement based process models' parameter
- CFD based mass and heat transfer analysis
- CFD-FEM framework
- Binary search algorithm based data transfer scheme
- Heating rate dependent austenization critical temperature
- New concept of thermal-metallurgical strain behavior

Also, the recent CFD-based process analyses and results that can be extended to multi-physical analysis were briefly introduced. By using the combination of suggested T-M-Me analysis method and CFD welding analysis, it is possible to reproduce a phenomenon closer to reality. However, considerable assumptions and simplified models have differences from real welding phenomena. To solve this gap and to use welding simulation as a prediction tool rather than a reproduction, many young researchers will need to challenge.

Acknowledgements The authors gratefully acknowledge the support of the Brain Korea 21 plus program and Mid-career Researcher Program through NRF of Korea (2013R1A2A1A01015605).

References

1. Rohde J, Jeppsson A (2000) Literature review of heat treatment simulations with respect to phase transformation, residual stresses and distortion. Scand J Metall 29(2):47–62
2. Cho SH, Kim JW (2002) Analysis of residual stress in carbon steel weldment incorporating phase transformations. Sci Technol Weld Joining 7(4):212–216
3. Deng D (2009) FEM prediction of welding residual stress and distortion in carbon steel considering phase transformation effects. Mater Des 30(2):359–366
4. Taljat B, Radhakrishnan B, Zacharia T (1998) Numerical analysis of GTA welding process with emphasis on post-solidification phase transformation effects on residual stresses. Mater Sci Eng A 246(1):45–54
5. Yaghi A, Hyde T, Becker A et al (2008) Finite element simulation of welding and residual stresses in a P91 steel pipe incorporating solid-state phase transformation and post-weld heat treatment. J Strain Anal Eng Des 43(5):275–293
6. Tsirkas SA, Papanikos P, Kermanidis T (2003) Numerical simulation of the laser welding process in butt-joint specimens. J Mater Process Technol 134:59–69
7. Goldak J, Chakravarti A, Bibby M (1984) A new finite element model for welding heat sources. Metall Trans B 15(2):299–305
8. Cho YT, Na SJ (2005) Application of Abel inversion in real-time calculations for circularly and elliptically symmetric radiation sources. Meas Sci Technol 16(3):878–884
9. Cho DW, Lee SH, Na SJ (2013) Characterization of welding arc and weld pool formation in vacuum gas hollow tungsten arc welding. J Mater Process Technol 213(2):143–152

10. Deng D, Murakawa H (2006) Prediction of welding residual stress in multi-pass butt-welded modified 9Cr–1Mo steel pipe considering phase transformation effects. Comput Mater Sci 37 (3):209–219
11. Miyao K, Wang Z, Inoue T (1986) Analysis of temperature, stress and metallic structure in carburized-quenched gear considering transformation plasticity. J Soc Mat Sci Jpn 35 (399):1352–1357
12. Nagasaka Y, Brimacombe J, Hawbolt E et al (1993) Mathematical model of phase transformations and elastoplastic stress in the water spray quenching of steel bars. Metall Trans A 24(4):795–808
13. Kang SH, Im YT (2007) Three-dimensional thermo-elastic–plastic finite element modeling of quenching process of plain-carbon steel in couple with phase transformation. Int J Mech Sci 49(4):423–439
14. Yang QX, Yao M, Park JK (2004) Numerical simulations and measurements of temperature and stress field in medium-high carbon steel specimen after hard-face-welding. Comput Mater Sci 29(1):37–42
15. Deng D, Murakawa H (2008) Finite element analysis of temperature field, microstructure and residual stress in multi-pass butt-welded 2.25 Cr–1Mo steel pipes. Comput Mater Sci 43 (4):681–695
16. Deng D, Murakawa H (2013) Influence of transformation induced plasticity on simulated results of welding residual stress in low temperature transformation steel. Comput Mater Sci 78:55–62
17. Fischer F, Oberaigner E, Tanaka K et al (1998) Transformation induced plasticity revised an updated formulation. Int J Solids Struct 35(18):2209–2227
18. Fischer F, Reisner G, Werner E et al (2000) A new view on transformation induced plasticity (TRIP). Int J Plast 16(7):723–748
19. Cheon J, Kiran DV, Na SJ (2016) Thermal metallurgical analysis of GMA welded AH36 steel using CFD–FEM framework. Mater Des 91:230–241
20. Cheon J, Na SJ (2016) Influence of simulation methods of temperature distribution on thermal and metallurgical characteristics in GMA welding. Mater Des 108:183–194
21. Cheon J, Na SJ (2017) Prediction of welding residual stress with real-time phase transformation by CFD thermal analysis. Int J Mech Sci 131(132):37–51
22. Standard ASTM E407 (2007) Standard practice for microetching metals and alloys. ASTM International. https://doi.org/10.1520/E0407-07
23. Hirt C, Nichols B (1988) Flow-3D user's manual. Flow Science Inc
24. Cho MH, Lim YC, Farson DF (2006) Simulation of weld pool dynamics in the stationary pulsed gas metal arc welding process and final weld shape. Weld J 85(12):271–283
25. Kikuchi N (1986) Finite element methods in mechanics. CUP Archive
26. Patankar S (1980) Numerical heat transfer and fluid flow. CRC Press
27. Cormen TH, Leiserson CE, Rivest RL et al (2001) Introduction to algorithms. MIT Press, Cambridge
28. Oliveira FLG, Andrade MS, Cota AB (2007) Kinetics of austenite formation during continuous heating in a low carbon steel. Mater Charact 58(3):256–261
29. Macedo MQ, Cota AB, Araújo FGDS (2011) The kinetics of austenite formation at high heating rates. Rem—Revista Escola de Minas 64(2):163–167
30. Cho DW, Na SJ, Cho MH et al (2013) Simulations of weld pool dynamics in V-groove GTA and GMA welding. Weld World 57(2):223–233
31. Cho DW, Na SJ (2015) Molten pool behaviors for second pass V-groove GMAW. Int J Heat Mass Transf 88:945–956
32. Wu L, Cheon J, Kiran DV et al (2016) CFD simulations of GMA welding of horizontal fillet joints based on coordinate rotation of arc models. J Mater Process Technol 231:221–238
33. Cho DW, Song WH, Cho MH et al (2013) Analysis of submerged arc welding process by three-dimensional computational fluid dynamics simulations. J Mater Process Technol 213 (12):2278–2291

34. Kiran DV, Cho DW, Song WH et al (2014) Arc behavior in two wire tandem submerged arc welding. J Mater Process Technol 214(8):1546–1556
35. Kiran DV, Cho DW, Song WH et al (2015) Arc interaction and molten pool behavior in the three wire submerged arc welding process. Int J Heat Mass Transf 87:327–340
36. Cho DW, Kiran DV, Na SJ (2017) Analysis of molten pool behavior by flux-wall guided metal transfer in low-current submerged arc welding process. Int J Heat Mass Transf 110:104–112
37. Cho DW, Na SJ, Cho MH et al (2013) A study on V-groove GMAW for various welding positions. J Mater Process Technol 213(9):1640–1652
38. Cheon J, Kiran DV, Na SJ (2016) CFD based visualization of the finger shaped evolution in the gas metal arc welding process. Int J Heat Mass Transf 97:1–14
39. Ahn JS, Na SJ (2013) Three-dimensional thermal simulation of nanosecond laser ablation for semitransparent material. Appl Surf Sci 283:115–127
40. Han SW, Ahn JS, Na SJ (2016) A study on ray tracing method for CFD simulations of laser keyhole welding: progressive search method. Weld World 60(2):247–258
41. Han SW, Cho WI, Na SJ et al (2013) Influence of driving forces on weld pool dynamics in GTA and laser welding. Weld World 57(2):257–264
42. Zhang YX, Han SW, Cheon J et al (2017) Effect of joint gap on bead formation in laser butt welding of stainless steel. J Mater Process Technol 249:274–284
43. Ge W, Han SW, Fang Y et al (2017) Mechanism of surface morphology in electron beam melting of Ti6Al4 V based on computational flow patterns. Appl Surf Sci 419:150–158
44. Cho DW, Cho WI, Na SJ (2014) Modeling and simulation of arc: laser and hybrid welding process. J Manufact Process 16(1):26–55

Liquid Metal Embrittlement of Galvanized Steels During Industrial Processing: A Review

Zhanxiang Ling, Min Wang and Liang Kong

Abstract Liquid metal embrittlement is the cause of reduction of elongation to failure and early fracture if normally ductile metals or alloys are stressed while in contact with liquid metals. Scientists have confirmed that many solid steel-liquid metal couples are subject to liquid metal embrittlement, one of them is solid steel-liquid zinc. Due to the wide use of zinc-coated galvanized steels, this couple has drawn much attention. This paper briefly introduces liquid metal embrittlement, with emphasis on the solid steel-liquid zinc couple and its occurrence in the process of industrial production in the literature. We first reviewed the findings that galvanized steels suffer embrittlement during experimental hot tensile test to understand its fundamental characteristics. We then summarized the occurrence of liquid metal embrittlement in galvanized steels during industrial processing, such as hot-dip galvanizing, hot stamping and welding.

Keywords Liquid metal embrittlement · Galvanized steel · Hot tensile test
Hot stamping · Welding

1 Introduction

Liquid metal embrittlement (LME), also known as liquid metal induced embrittlement (LMIE) or liquid metal assist cracking (LMAC), is the reduction on elongation to failure and early fracture if normally ductile metals or alloys are stressed while in contact with liquid metals [1]. Cracking arises in LME is classified as one case of environmental assist cracking (EAC). Although causing a lot of

Z. Ling · M. Wang (✉) · L. Kong
Shanghai Key Laboratory of Materials Laser Processing and Modification,
Shanghai Jiao Tong University, Shanghai 200240, China
e-mail: wang-ellen@sjtu.edu.cn

Z. Ling · M. Wang
The State Key Laboratory of Metal Matrix Composites, Shanghai Jiao Tong University,
Shanghai 200240, China

© Springer Nature Singapore Pte Ltd. 2018
S. Chen et al. (eds.), *Transactions on Intelligent Welding Manufacturing*,
Transactions on Intelligent Welding Manufacturing,
https://doi.org/10.1007/978-981-10-8330-3_2

troubles in practice, the phenomenon is not fully understood compared with some other EAC, such as hydrogen embrittlement and stress corrosion cracking. According to a review paper [2], the earliest study of LME was published in 1914, in which the internal stressed brass was fractured when in contact with mercury was discussed. Up until now, LME has been studied for over a century, the existing scientific paper which concerns the susceptibility of various specific combinations to LME is plentiful but dispersive [3]. Scientists have been dedicated to find out a universal mechanism to explain all the LME phenomenon and predict its occurrence, and several modes have been proposed [1–3]. The most recent contribution was done by Bauer et al. [4], in which they combined density functional theory calculations with thermodynamic considerations to investigate the LME of iron by liquid zinc. Nevertheless, none of them can fully account for all experimental observations.

Generally, several features are widely accepted for LME:

(a) Particular solid metal-liquid metal couples are prone to LME, which means some couples are susceptible to embrittlement while others appear to be immune. This is referred to as the specificity of LME.
(b) A critical stress is required for the occurrence of LME, and the liquid metal must direct contact on an atomic scale with the stressed solid metal.
(c) LME results in initial intergranular and brittle fracture in most cases for the polycrystalline metal and alloy.
(d) A specific brittle-ductile transition temperature always exists in LME for different embrittling couples.

Among all the materials, steels own a dominant position for industrial use due to their abundant source, economic efficiency and comprehensive performance. Thus, numerous studies focused on the LME of various steels and they are proved to have a poor resistance to LME. Due to the serious condition which satisfies the prerequisites, LME of steels is frequently reported under the circumstance of nuclear applications, as they can be embrittled by liquid Pb-17Li [5], Pb [6], or Pb–Bi eutectic [7] in service. Under experimental condition, steels can also be embrittled by Cu [8], Na [9], Sn [10] and so on. Recently, a clutch shell of a motorbike made from SPCC nitrided steel [11] and turbine casing segment screws of an aeroengine made from 35NC6 steel [12] are reported to suffer LME, indicating LME phenomena extensively exist in actual use.

Empirical rules suggest that an embrittlement couple may have limited mutual solubility and low tendency to form stable intermetallic compounds (IMCs) [3], in contrast with the rules, zinc can embrittle steel, even if they have relatively good mutual solubility and can form stable IMCs, as can be seen in Fig. 1. For over 200 years, the zinc coating is widely used to provide corrosion resistance for steel, it can separate the steel from the corrosion environment and act as the sacrificial anode [14]. From the perspective of LME, zinc coating provides the potential liquid metal film so that LME may occur in steel if the temperature and stress conditions meet the prerequisites of LME. The melting point of zinc is low (about 419 °C),

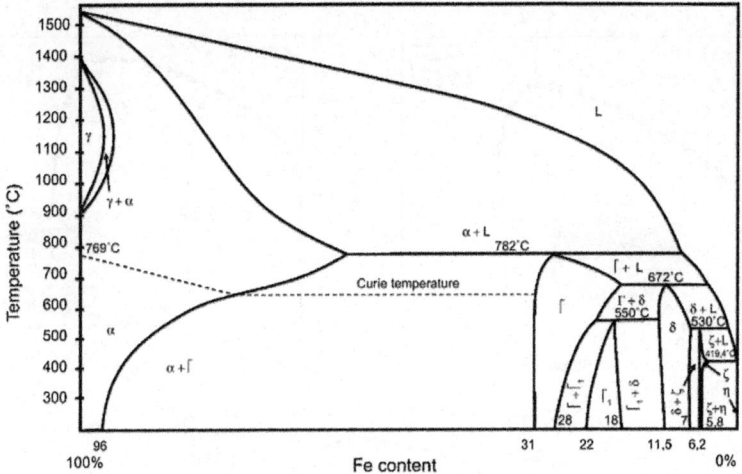

Fig. 1 Iron–Zinc binary diagram [13]

and as expected, galvanized steels are susceptible to LME cracking phenomena when they are subjected to some hot industrial processes such as hot-dip galvanizing, hot stamping and welding according to opening literatures. This paper gives a brief review of these literatures and try to offer a comprehensively understanding of LME phenomena of galvanized steels during industrial processes.

2 LME of Galvanized Steels During Hot Tensile Test

When doing research on LME of various embrittlement couples, materials scientists always carry out hot tensile test to obtain the data characterizing the phenomena, and nowadays, hot tensile tests are often performed using a Gleeble thermo-mechanical simulator. The experiment can provide ideal and quantitative conditions in laboratory, while stressed specimens are in contact with liquid metal under certain strain rate and temperature. Thus, for the solid steel-liquid zinc combination, this test method is also frequently used. Some of the achievements about LME of steel by zinc were reported in Japanese language papers many years ago [15–19]. In this section, a few recent papers using hot tensile test to study liquid zinc embrittlement on steels are introduced previously to help understand the embrittlement phenomenon. Several factors were found to have influence on the LME.

Beal et al. [20–22] studied the LME of electrogalvanized (EG) twinning-induced plasticity (TWIP) steel, the steel is fully austenitic at room temperature, and the contact with liquid zinc was due to the melting of the zinc coating. Figure 2 shows the tensile curves of uncoated and EG steels at different temperatures under a strain

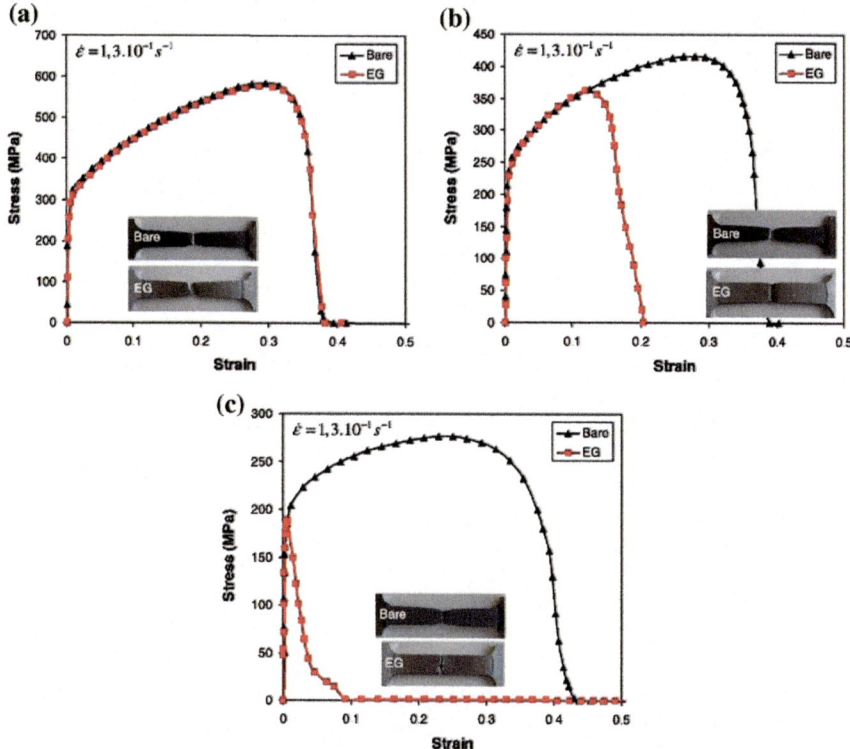

Fig. 2 Tensile curves of bare and electrogalvanized (EG) specimens obtained at different temperatures: **a** 600 °C, **b** 700 °C and **c** 800 °C [22]

rate of 0.13/s. It can be seen that the EG specimen was not embrittled at 600 °C, but its ultimate tensile strength and fracture elongation reduce compared with uncoated one at 700 °C. The steel is more severely embrittled at 800 °C, as fracture occurs at a very small strain. It is worth noticing that the lower embrittlement temperature is much higher than the melting point of zinc, it may attribute to the experimental procedure where liquid zinc came from the melt of zinc coating, the fast process required higher temperature for zinc coating to be fully melted. It seems convincing as LME of steel by zinc were found to occur at about 450 °C in liquid zinc bath during galvanizing [23]. However, Barthelmie et al. [24] reported that the galvanized surface refined TWIP steel could be embrittled at 450 °C during hot tensile test. Frappier et al. [25] investigated the embrittlement of an EG advanced multiphase high strength steel which consists of ferrite/bainite matrix with about 12% of retained austenite, the test conditions were similar with Beal et al. and they obtained same LME starting temperature. They claimed that the beginning of

Fig. 3 Influence of strain rate on the embrittlement of the Fe22Mn0.6C steel by liquid zinc [21]

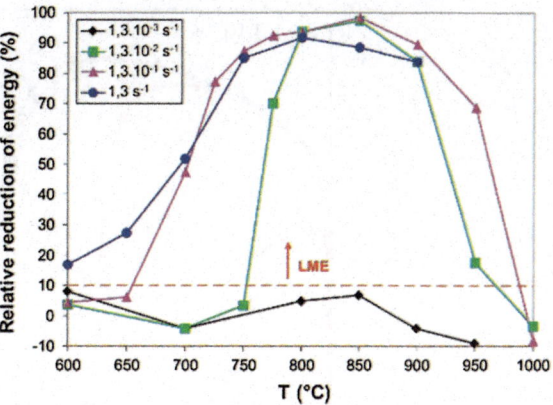

embrittlement at 700 °C was due to a wetting transition, the liquid metal can wet the grain boundary (GB) if

$$\gamma_{GB} > 2\gamma_{SL} \qquad (1)$$

where γ_{GB} is the grain boundary energy and γ_{SL} is the solid steel-liquid zinc interfacial energy. Microscopic image of steel/Zn interface gave the evidence of GB wetting by Zn, so they believed the lower temperature for LME was the wetting transition temperature for steel/Zn couple, which explained why LME didn't occur at melting point of zinc. Figure 3 demonstrates the variation of relative reduction of energy as a function of the temperature for the four strain rates, it clearly indicates that at low strain rate (black line), no significant LME occurs at all test temperatures, and the higher the strain rate, the lower temperature at which the LME occurs. The graph also illustrates the "ductility trough" and "brittle-ductile transition" features which are commonly observed in many LME systems. The ductility of the steel was lost within a temperature range of 700–950 °C, and fully regained at about 1000 °C. Beal et al. [21] hypothesized that the recovery of ductility at high temperature is due to the evaporation of zinc, whose boiling point is about 907 °C, which makes the liquid zinc insufficient for embrittlement to occur. Figure 4 presents the influence of holding time in the LME, the specimens were pre-exposed to liquid zinc and maintained for a certain holding time before the tensile test. It can be seen that long holding time leads to a recovery of ductility, and when the holding time increases to 20 s, the tensile curve of EG specimen is highly coincident with uncoated one and the embrittlement is fully suppressed. Beal et al. [22] claimed that the formation of IMCs at the interface between steel and zinc was responsible for the recovery when increasing the holding time, as they prevented the contact of steel with liquid zinc.

Jung et al. [26] investigated the influence of constituent microstructure of the galvanized steel and pre-strain on the LME, three kinds of steels, i.e. deep draw quality (DQ) steel which consists of fully ferritic, dual-phase (DP) steel which

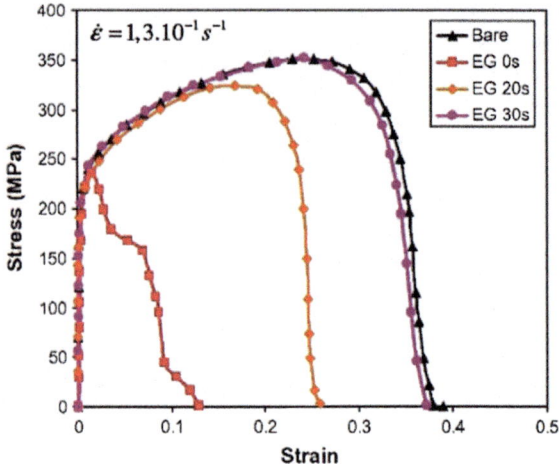

Fig. 4 Tensile curves obtained at 750 °C (strain rate 0.13/s): progressive ductility recovery after holding at 750 °C [22]

Table 1 Summarized results of hot tensile test to reveal LME occurring conditions of each alloy (orange colored box = ferrite, blue colored box = austenite) [26]

	Strain rate (s⁻¹)	600 °C	700 °C	800 °C	900 °C
	1			LME	LME
DQ	0.1			LME	LME
	0.01			LME	LME
	1	LME	LME	LME	LME
DP	0.1		LME	LME	LME
	0.01			LME	LME
	1		LME	LME	LME
TWIP	0.1		LME	LME	LME
	0.01			LME	LME

consists of ferrite and martensite, TWIP steel which consists of austenite, were chosen to carry out the hot tensile test. The overall results on the occurrence of LME were summarized in Table 1. It proves that the occurrence of LME is irrelevant to the original microstructure of steel. Results also show that the temperature and strain rate have an influence on the LME, which is consistent with previous statements. Furthermore, under the conditions of temperature of 600 °C and strain rate of 0.1/s, no LME occurred, but DP and TWIP steels were embrittled under the same conditions after they were pre-strained up to 0.4% at 900 °C. Taking into account that the engineering strain of 0.4% just passed the yield point, the authors claimed that not only stress but also plastic deformation was needed for the occurrence of LME. With the help of TEM images and EDS line profiling results, they found that the pre-strain could accelerate zinc diffusion into substrate grain boundaries (GBs) as well as weakened them and subsequently caused the LME even at the conditions where LME normally did not occur.

Kang et al. [27] researched the LME of a Zn-coated Interstitial-Free (IF) steel, a Zn-coated 22MnB5 press-hardened steel (PHS) and a Zn-coated TWIP steel, and found that the PHS and TWIP steel suffered LME at 850 °C. But no LME occurred in IF steel at both 850 °C (below A_{c1} temperature) and 950 °C (above A_{c3} temperature), this indicated that the LME of steels was not directly related to their crystal structure but the composition and type of steel did influence the LME process. The LME occurred due to the penetration of Zn, and the average penetration depth of the IF steel, PHS and TWIP steel were 53, 152 and 303 μm respectively, which explained their differences in the LME behavior. The Zn percolation along GBs and rapid solid-state Zn grain boundary diffusion were proposed to be compatible with the Zn penetration process.

From the above statements, it can be concluded that the temperature, strain rate, pre-strain, and type of steel all influence the LME of steel by liquid zinc in different ways. Particular conditions needed for LME to occur can be experienced during hot working processes such as hot-dip galvanizing, hot stamping and welding. These have been confirmed by many researches and they will be discussed next.

3 LME of Steels During Hot-Dip Galvanizing

Zinc coatings are predominantly used to improve the corrosion resistance of steel, typical processing methods used in producing zinc coatings include hot-dip galvanizing (GI), galvannealing (GA) and electrogalvanizing (EG). For hot-dip galvanizing method, the steels are dipped into 445–455 °C molten zinc bath and the immersion times are in the range of 3–6 min [28]. Before hot-dip galvanizing, the steel surface is carefully cleaned to remove any impurities so that the steel substrate is directly in contact with liquid zinc during the hot-dip galvanizing process. LME cracking sometimes appear during hot-dip galvanizing in large structural steel components like beams and profiles, or welded structures, as shown in Fig. 5. Mraz and Lesay [29] concluded that the stress needed for the cracking came from local residual stresses as the consequences of welding and local strains as the

Fig. 5 Observation of LMAC in a steel structure after hot-dip galvanizing [13]

consequences of heating during galvanizing. Under the situation, James [30] suggested that a good design of structural steelworks can help against the LME during galvanizing.

To investigate the LME phenomena during galvanizing, scientists always apply external load to better present the results. Carpio et al. [31] researched how environmental factors acted in steel embrittlement during galvanizing, the structural S450J0 steel was chosen as the main study material, its low ductility and high strength made it very prone to suffer failure during galvanizing. Tensile and Charpy impact test results showed that the properties of studied steels dropped at 450 °C compared with that at room temperature, so the steel was softer and more brittle at galvanizing temperature, but it was not the main cause of the embrittlement during galvanizing. Fluxing treatment are always used to prevent oxidation before galvanizing. The study indicated that the fluxing increased the surface roughness so as to enhance the susceptibility of local embrittlement due to notch effect, which was related to stress concentration. Fluxing as well as galvanizing led to the hydrogen accumulation, it gave a possibility that the embrittlement might be caused by hydrogen. This hypothesis was denied also by Carpio et al. [32] in another paper, in which they tested and calculated the hydrogen concentration in the steel base and zinc layer and found that the hydrogen mainly presented in the zinc layer and hardly existed in the steel base. Nevertheless, Mraz and Lesay [29] complained that the hydrogen embrittlement was responsible for crack initiation and LME was responsible for crack propagation as they observed the transgranular fracture at crack initiation sites and intergranular fracture at crack propagation sites. J toughness tests were carried out on compact tensile specimens at 450 °C in air and two different Zn baths: traditional Zn–Pb bath and innovative Zn–Pb–Sn–Bi bath [31]. Results showed that the toughness of the samples further decreased in Zn bath and the embrittlement was more aggressive in the Zn–Pb–Sn–Bi bath, as shown in Fig. 6. It was because Sn and Pb were accumulated next to the steel base and

Fig. 6 Toughness J-test results on CT specimens made of S450J0 steel in two liquid Zn baths: Traditional (Zn–Pb) and Zn–Pb–Sn–Bi baths [31]

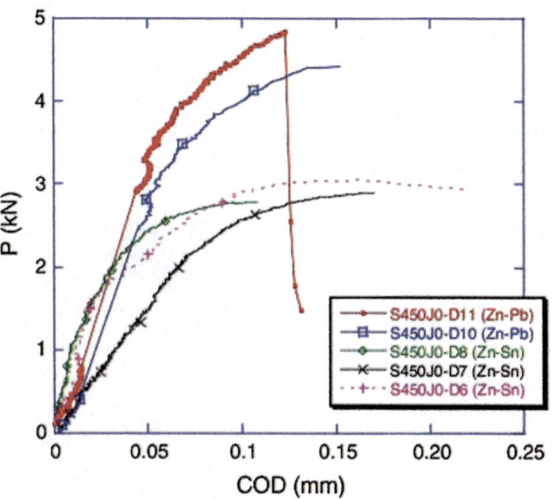

formed low-melting-point eutectics (about 180 °C, even less if Bi was present), they flowed easily to the cracks and were very reactive with the steel, thus facilitated the embrittlement process.

Mendala [23] applied tension stretching with different levels (400–800 MPa) to the C70D steel during hot dip galvanizing at 450 °C. Two liquid baths were used, i.e. pure zinc and zinc with 2% tin addition. Results showed that if the load was applied to the samples and suddenly released, no cracking occurred for all samples under different load values. If the constant load was applied to the samples during galvanizing, no cracking occurred in the samples that dipped in a zinc bath but cracking was detected in the samples during metallization in a zinc bath with 2% tin addition under high stress values, 600 and 700 MPa, and the sample was ruptured under 800 MPa. The author deduced that the constant load could let the internal stresses accumulate and the cracks would occur in liquid bath as a result of loss of cohesive properties. The explanation of the conducive effect of tin was not given in the paper, but as tin is a severe steel embrittler and the embrittlement can happen at 266 °C [10], the concept of "Insert Carriers" [3] may be introduced to explain the phenomenon.

The influence of previous cold deformation on LME of typical structural steel S235JR in 450 °C liquid zinc bath was studied by Luithle and Pohl [33]. The samples were firstly tested in 450 °C hot air, their tensile strength increased and reduction of area decreased with the increase of deformation degree owing to the work-hardening. Thus, the severity of LME was evaluated by the ratio of reduction in area of samples tested in 450 °C liquid zinc bath and 450 °C hot air under the same deformation degree. Results showed that the severity of LME decreased with the increasing degree of deformation, and the fracture surfaces changed from intergranular cleavages to ductile dimples. If the deformation was large enough, there were almost no difference between the samples tested in hot air and liquid zinc. The authors explained that with increasing cold deformation the grains became more and more stretched in axial direction, the original GBs (preferred intergranular crack paths) which were perpendicular to the applied force became more parallel to the load, and thus, the component stress was not enough to open the GB, leading to the removal of LME.

4 LME of Galvanized Steels During Hot Stamping

Hot stamping, also called press hardening, was developed in accordance with the demand for ultra-high strength steels in automobile industry. Currently, there are two different hot stamping methods, i.e. the direct hot stamping process in which a blank is heated up in a furnace, transferred to the press and subsequently formed and quenched in the closed tool, and the indirect process characterized by the use of a nearly complete cold pre-formed part which is subjected only to a quenching and calibration operation in the press after austenitization [34]. To protect the steels from oxidation and provide cathodic corrosion protection during hot stamping, the steels are often galvanized before the process. However, Zn-coated steels are easily

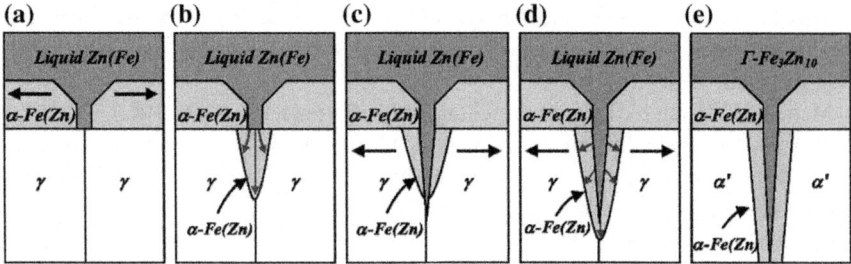

Fig. 7 Schematic illustrating the mechanism of Zn grain boundary diffusion-mitigated phase transformation leading to crack formation on Zn-coated PHS during die quenching (γ: austenite, α': martensite) [37]

subjected to LME cracks during hot stamping process, especially the direct hot stamping process [35], which has potentially bad effect on the mechanical properties of the parts.

Lee and his coworkers published several papers about the LME of galvanized 22MnB5 during direct hot stamping simulated by a Gleeble 3500 thermo-mechanical process simulator. In one paper published in 2012 [36], they found that when the specimen was deformed at 850 °C, Zn penetrated into the steel matrix and caused brittle fracture due to grain boundary decohesion. The fracture didn't occur when the specimen was held at 850 °C for 4 min, quenched to 700 °C and deformed at this temperature. It was because that the solid Γ_1 intermetallic compound was formed as a result of a peritectic reaction between solid α-Fe and liquid Zn at 782 °C during quenching, and the absence of liquid Zn inhibited the LME. Increasing the annealing time, i.e. soaking at 850 °C to 20 min before hot stamping also suppressed the LME as the coating layer was fully transformed to α-Fe (Zn), no liquid Zn presented to cause LME in this case. In another paper published in 2014 [37], they gave a detailed mechanism for LME cracking during hot stamping, as illustrated in Fig. 7: (a) High-temperature crack initiation at an α-Fe(Zn) grain boundary in the surface alloy layer; (b) Zn diffusion along the γ grain boundary and transformation of the Zn-diffused γ grain boundary region to α-Fe(Zn); (c) Crack propagation through the weak α-Fe(Zn) grain boundary layer; (d) Crack propagation by repetition of the diffusion-transformation stages (b) and (c); (e) After cooling, the high-temperature Zn_{liq} distribution is reflected in the room-temperature distribution of Γ-Fe3Zn10. The absence of transformation of γ to α' lath martensite allows for the identification of the Zn diffusion layer in the vicinity of the crack tip. Based on the model, they postulated that LME crack was caused not by liquid Zn, but by the presence of a thin α-Fe (Zn) layer at austenite grain boundaries formed by the Zn diffusion-mitigated phase transformation of the boundary region. The strength of this ferrite layer was low compared with the austenite, leading to the fracture during hot stamping. In the most recent work, Lee et al. [38] carried out tensile tests and three-point bending tests on the PHS after hot stamping. Results showed that the Zn coating has no cacoethic influence on tensile properties regardless tested in the length direction (LD) or transverse direction (TD) after hot stamping, but it deteriorated the bending performance of wall side significantly due to the LME cracking in the location. TD-oriented samples provided the worst bendability because the bending direction coincided with the microcracks propagation direction. Lee et al. [39, 40] also studied the effect of a 55 wt% Al–Zn coating on PHS during hot stamping and found the steel was not susceptible to LME, it was due to the fact that the liquid Zn was fully confined to the Al–Zn layer as intergranular islands or at the Fe–Al intermetallic grain boundaries, as the solubility of the Zn in the Fe–Al compounds was very low.

In the research by Drillet et al. [41], the cracks formed in the steel during hot forming were classified to macro-cracks (>100 μm) and micro-cracks. The macro-cracks were formed due to the liquid Zn penetration in the steel grain boundaries under stress, i.e. the LME. The cracks often located on the external side of the radius where the steel was under tensile stress. In this case, they claimed that the GI steel was only dedicated to indirect hot stamping process, but the GA steel could be dedicated to both direct and indirect hot stamping process with suitable heat treatment. The micro-cracks always initiated in the wall, where the friction between the steel sheet and the tools was very high. Kurz et al. [42, 43] also claimed that to avoid the LME during hot stamping, the indirect process was in industrial application for galvanized steels these days, however, it led to a much higher production cost. They introduced the so-called direct process with pre-cooling and the modified PHS, 20MnB8 and successfully avoided the LME cracking. Seok et al. [44] suggested that during the direct process, the heating time of 5 and 10 min with a heating temperature of 850 °C, heating times of 5 and 10 min with a heating temperature of 900 °C, and a heating time of 3 min with a heating temperature of 950 °C were appropriate for the product to minimize the LME cracks. Way to apply the direct hot forming process to galvanized steels and avoid the LME cracking can also be found in a patent [45].

5 LME of Galvanized Steels During Welding

The most dominant joining method in manufacturing industry is welding. During the welding process, the peak temperature in welding zone and heat affected zone (HAZ) is very high and far beyond the melting point of Zn. For the galvanized steel, Zn exists in liquid form and presents on the surface of solid steel in HAZ during fusion welding, thus there is a high risk of LME if the stress condition was up to grade in this area.

LME cracking is frequently observed in galvanized steels during resistance spot welding (RSW) [46–50]. Although some reports showed that the LME cracks have no significant influence on the tensile and fatigue properties of RSW joints [45, 51], but the surface crack is a potential threat to the performance and integrity of structures.

Kim et al. [52] detected surface LME cracks in RSW joints of the Zn-coated transformation-induced plasticity (TRIP) steel. The cracks mainly located in the concaves of welding centers and the inclined regions. It was found that welding force, welding current and welding time all had significant effect on surface cracking, it increased with the increase of welding current and welding time and the decrease of welding force. The holding time had less effect than other factors, the increase of holding time slightly decreased surface cracking. The electrode type also influenced the location and number of crack. By SEM observation and EDS analysis, they found that a Cu_5Zn_8 IMC, which formed by alloying with the Cu electrode, was present on the crack surface. An improvement method in this study was introduced, i.e. using a pre-current of 10 kA and 3 cycles to melt the Zn layer and a cooling time of 6 cycles to facilitate the removal of molten Zn before the application of welding current. Due to the lack of Zn, LME cracking was suppressed.

Barthelmie et al. [53] researched dissimilar RSW of the galvanized TWIP steel to the galvanized HX340LAD steel, they detected LME cracks on the TWIP steel side but not on the HX340LAD steel side, and they attributed it to the austenitic structures' sensitivity to LME. They also investigated the influence factors of the LME and found that the smaller heat input, the larger electrode force and electrode cap diameter could decrease the LME crack length.

Tolf et al. [54] found that the coating type and wear degree of the electrode cap had remarkable effect on LME cracking during RSW. The welds of GI coated dual phase (DP) steel were more prone to surface cracking compared with EG coated DP steel. With GI coating, cracks were observed when welding the first sample, and further increase in number and crack length with the ongoing of welding process. However, with EG coating, the first 50 welds were crack free. The authors claimed that the small amount of Al in hot dip galvanized coating was the key factor, as it was oxidized and forming aluminum oxide on the steel surface, and significantly increased the resistance during RSW. Higher resistance resulted in higher heat generation, and then higher LME cracking susceptibility.

Ashiri et al. [55] built up the concept of "supercritical area" and "critical nugget diameter" to describe the LME phenomena during RSW of Zn-coated TWIP steels. They found that most of the cracks were formed in the periphery area at the vicinity of the contacted area between the electrodes and TWIP steel sheets, as shown in Fig. 8. They defined the peripheral area as "supercritical LME area". SORPAS simulations confirmed that this area experienced highest temperature and stress, which gave an explanation that the conditions in this area was favorable for LME. In addition, the cooling condition in this area was worst because there was a gap between the electrode and this area. The authors further discovered that there was a lowest nugget diameter for LME to occur, which were 6.15 mm for GI coated steel, 6.21 mm for GA coated steels and 6.42 mm for EG coated steel, thus the EG coated steel had the lowest LME susceptibility. The maximum crack length increased with the increase of nugget diameter. They defined the lowest nugget diameter as "critical nugget diameter", which could be correlated with the critical tensile stress and temperature required for LME to occur. In another paper published by Ashiri et al. [56], the authors developed a smart welding procedure to produce LME-free welds of Zn-coated TWIP steel at high temperature. In their method as shown in Fig. 9, a two-pulse base current which was set to 5 kA provided the first step of heat input required to form minimum nugget diameter, then the second pulse current

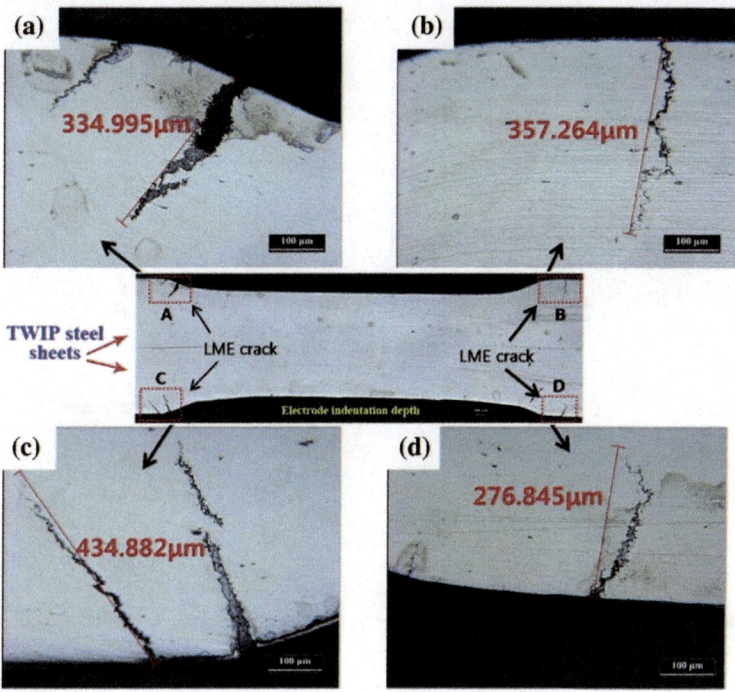

Fig. 8 Observation of LME in a resistance spot welded TWIP steel [55]

could be increased to the current where LME occurred. In the best welding schedule as shown in Fig. 9c, the weldable current rage without LME cracks was 85.7% greater than the current rage in single-pulse welding schedule. The LME cracking didn't occur until the expulsion, the expulsion was visible and easy to avoid, thus LME-free welds were obtained. SORPAS simulation results indicated that the temperature and stress concentration at the supercritical LME area of the impulse welded sample was much lower than that of the single-pulse welded one, which provided less liquid zinc and less tensile stress for LME to occur.

LME can also occur during arc welding. Bruscato [57] detected LME cracks when welded austenitic stainless steel to galvanized steel, and suggested that the zinc coating must be scrupulously removed from the joint area prior to welding to preclude LME cracking. Mori and Nishimoto [58] reported that LME intergranular cracking sometimes occurred in HAZ of dissimilar welded joints of austenitic stainless steels with galvanized carbon steels, the susceptibility of LME differed with various chromium and nickel contents which was due to the change in grain boundary energy in the austenitic steels. Pańcikiewicz et al. [59] also observed LME cracking in T-joints welded by Gas Metal Arc Welding (GMAW) between a hot dip galvanized E275D steel and a AISI 304 stainless steel. The cracks located in the HAZ of the stainless steel with length up to 3 mm, and had an intergranular character, Zn atoms presented on the crack surface. The zinc melted on the

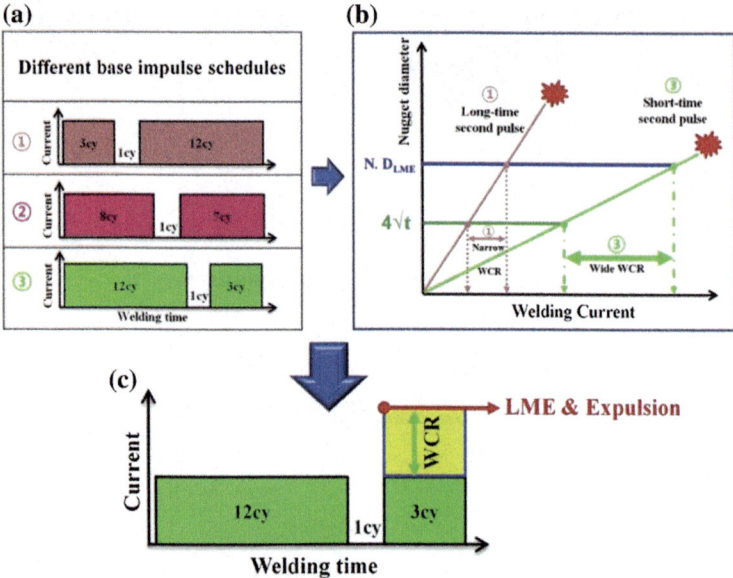

Fig. 9 **a** Different base impulse schedules; **b** the criterion for the selection and **c** conditions of the best welding schedule [56]

Fig. 10 A schema of the mechanism of zinc vaporizing and depositing on austenitic stainless steel during arc welding a T-joint with a fillet weld [59]

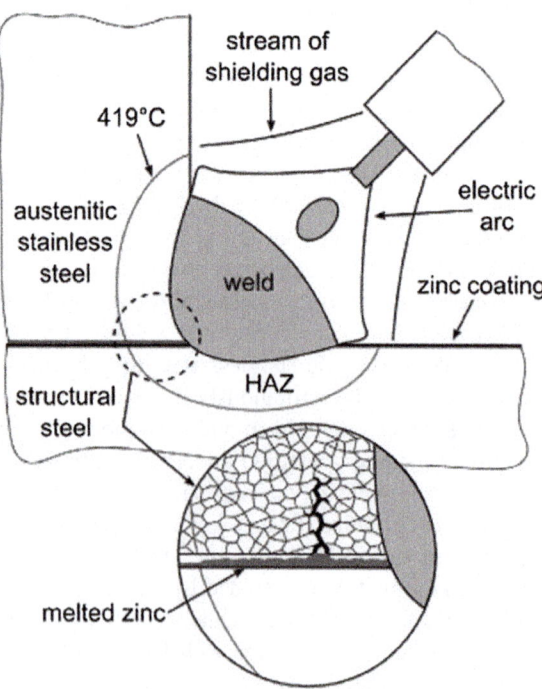

galvanized steel surface, then evaporated and condensed on the austenitic stainless steel by surface tension and adhesive forces. Liquid zinc slowly penetrated along grain boundaries, and the internal stress originated from phase transformation resulted in the fracture, as shown in Fig. 10.

6 Conclusion

Liquid zinc can embrittle various steels by LME mechanism. Many factors, such as temperature, strain rate, pre-exposure, cold deformation, coating types and constituents influence the severity of the embrittlement. The required conditions for LME may be reached during industrial processed such as hot-dip galvanizing, hot stamping and welding, the occurrence of LME cracking is an undesirable phenomenon and bring challenges to those industrial processes. Considering that the high temperature during the hot working industrial processes is inevitable, the stress condition should be carefully controlled to avoid the LME. Some attentions should be paid as follows:

1. High residual stress and stress concentration should not be reserved in the steelwork which is going to experience hot-dip galvanizing process, it can be achieved by rational design of large welded structures.
2. Try not to apply the galvanized steels to direct hot stamping process, and use the galvannealed steels to replace them.
3. The heat input during resistance spot welding of galvanized steels should be controlled by applying the novel welding procedure, electrogalvanized steels are good substitutes for hot-dip galvanized steels as they are able to reduce the risk of LME.

References

1. Fernandes PJL, Jones DRH (1997) Mechanisms of liquid metal induced embrittlement. Int Mater Rev 42(6):251–261
2. Nicholas MG, Old CF (1979) Liquid metal embrittlement. J Mater Sci 14(1):1–18
3. Kamdar MH (1983) Liquid metal embrittlement. Treatise Mater Sci Technol 25(1):361–459
4. Bauer KD, Todorova M, Hingerl K et al (2015) A first principles investigation of zinc induced embrittlement at grain boundaries in bcc iron. Acta Mater 90:69–76
5. Sample T, Fenici P, Kolbe H (1996) Liquid metal embrittlement susceptibility of welded MANET II (DIN 1.4914) in liquid Pb-17Li. J Nucl Mater 233:244–247
6. Legris A, Nicaise G, Vogt JB et al (2000) Embrittlement of a martensitic steel by liquid lead. Scripta Mater 43(11):997–1001
7. Hojna A, Di Gabriele F, Klecka J (2016) Characteristics and liquid metal embrittlement of the steel T91 in contact with lead–bismuth eutectic. J Nucl Mater 472:163–170
8. Padmanabhan B, Salunkhe P, Nage D (2015) Liquid metal embrittlement of austenitic stainless steel fitting caused by copper contamination. J Fail Anal Prev 15(4):480

9. Hémery S, Auger T, Courouau JL et al (2014) Liquid metal embrittlement of an austenitic stainless steel in liquid sodium. Corros Sci 83:1–5
10. Clegg RE, Jones DRH (2003) Liquid metal embrittlement of tensile specimens of En19 steel by tin. Eng Fail Anal 10(1):119–130
11. Ding N, Xu N, Guo W et al (2016) Liquid metal induced embrittlement of a nitrided clutch shell of a motorbike. Eng Fail Anal 61:54–61
12. Nandi V, Bhat RR, Yatisha IN et al (2012) Liquid-metal-induced embrittlement in turbine casing segment screws of an aeroengine. J Fail Anal Prev 12(4):348–353
13. Kuklik V, Kudlacek J (2016) Hot-dip galvanizing of steel structures. Butterworth-Heinemann, Boston, pp 30, 170
14. Marder AR (2000) The metallurgy of zinc-coated steel. Prog Mater Sci 45(3):191–271
15. Kikuchi M (1980) Liquid metal embrittlement of steels by liquid zinc. J Soc Mater Sci 29(317):181–186
16. Nakasa K, Takei H, Matsuda M (1988) Crack propagation behavior in liquid zinc embrittlement of mild steel. J Soc Mater Sci 37(413):166–170
17. Kikuchi M, Lezawa T (1982) Effect of stress-concentration factor on liquid metal embrittlement cracking of steel in molten zinc. J Soc Mater Sci 31(352):271–276
18. Kikuchi M (1981) Liquid metal embrittlement cracking of notched rectangular steel plate in molten zinc. J Soc Mater Sci 30(329):194–199
19. Nakasa K, Takei H, Takemoto S (1984) Effects of tensile speed, testing temperature and ferrite grain size on liquid zinc embrittlement in precracked specimens of mild steel. J Soc Mater Sci 33(372):1193–1198
20. Beal C, Kleber X, Fabregue D et al (2011) Liquid zinc embrittlement of a high-manganese-content TWIP steel. Philos Mag Lett 91(4):297–303
21. Beal C, Kleber X, Fabregue D et al (2012) Embrittlement of a zinc coated high manganese TWIP steel. Mater Sci Eng A 543:76–83
22. Beal C, Kleber X, Fabregue D et al (2012) Liquid zinc embrittlement of twinning-induced plasticity steel. Scripta Mater 66(12):1030–1033
23. Mendala J (2012) Liquid metal embrittlement of steel with galvanized coatings. IOP conference series-materials science and engineering, vol 35. IOP Publishing, Bristol, pp 1–8
24. Barthelmie J et al (2016) Liquid metal embrittlement in resistance spot welding and hot tensile tests of surface-refined TWIP steels. In: IOP conference series-materials science and engineering, vol 118. IOP Publishing, Bristol, pp 1–8
25. Frappier R et al (2014) Embrittlement of steels by liquid zinc: crack propagation after grain boundary wetting. In: Advanced materials research, vol 922. Trans Tech Publications, Zurich, pp 161–166
26. Jung G, Woo IS, Suh DW et al (2016) Liquid Zn assisted embrittlement of advanced high strength steels with different microstructures. Met Mater Int 22(2):187–195
27. Kang H, Cho L, Lee C et al (2016) Zn penetration in liquid metal embrittled TWIP steel. Metall Mater Trans A 47(6):2885–2905
28. Schulz WD, Thiele M (2011) Hot-dip galvanizing and layer-formation technology. Handb Hot-Dip Galvanization 91–124
29. Mraz L, Lesay J (2009) Problems with reliability and safety of hot dip galvanized steel structures. Soldagem & Inspecao 14(2):184–190
30. James MN (2009) Designing against LMAC in galvanised steel structures. Eng Fail Anal 16(4):1051–1061
31. Carpio J, Casado JA, Álvarez JA et al (2009) Environmental factors in failure during structural steel hot-dip galvanizing. Eng Fail Anal 16(2):585–595
32. Carpio J, Casado JA, Álvarez JA et al (2010) Stress corrosion cracking of structural steels immersed in hot-dip galvanizing baths. Eng Fail Anal 17(1):19–27
33. Luithle A, Pohl M (2015) On the influence of cold deformation on liquid metal embrittlement of a steel in a liquid zinc bath. Mater Corros 66(12):1491–1497
34. Karbasian H, Tekkaya AE (2010) A review on hot stamping. J Mater Process Technol 210(15):2103–2118

35. Feng GW et al (2016) Microcacks in galvannealed hot stamping 22MnB5 steel. In: Advanced high strength steel and press hardening-proceedings of the 2nd international conference. World Scientific, Changsha, pp 110–114
36. Lee CW, Fan DW, Sohn IR et al (2012) Liquid-metal-induced embrittlement of Zn-coated hot stamping steel. Metall Mater Trans A 43(13):5122–5127
37. Cho L, Kang H, Lee C et al (2014) Microstructure of liquid metal embrittlement cracks on Zn-coated 22MnB5 press-hardened steel. Scripta Mater 90:25–28
38. Lee CW, Choi WS, Cho L et al (2015) Liquid-metal-induced embrittlement related microcrack propagation on Zn-coated press hardening steel. ISIJ Int 55(1):264–271
39. Lee CW, De Cooman BC (2014) Microstructural evolution of the 55 Wt Pct Al-Zn coating during press hardening. Metall Mater Trans A 45(10):4499–4509
40. Lee CW, Choi WS, Cho YR et al (2015) Microstructure evolution of a 55wt.% Al–Zn coating on press hardening steel during rapid heating. Surf Coat Technol 281:35–43
41. Drillet P, Grigorieva R, Leuillier G et al (2013) Study of cracks propagation inside the steel on press hardened steel zinc based coatings. La Metallurgia Italiana 1:3–8
42. Kurz T, Luckeneder G, Manzenreiter T et al (2015) Zinc coated press-hardening steel-challenges and solutions. SAE technical paper No. 2015-01-0565
43. Kurz T, Larour P, Lackner J et al (2016) Press-hardening of zinc coated steel-characterization of a new material for a new process. In: IOP conference series-materials science and engineering, vol 159. IOP Publishing, Bristol, pp 1–16
44. Seok HH, Mun JC, Kang CG (2015) Micro-crack in zinc coating layer on boron steel sheet in hot deep drawing process. Int J Precis Eng Manuf 16(5):919–927
45. Sachdev AK, Brown TW (2015) Controlling liquid metal embrittlement in galvanized press-hardened components. US Patent Application 14/627,579
46. Zhang P, Xie J, Wang YX (2011) Effects of welding parameters on mechanical properties and microstructure of resistance spot welded DP600 joints. Sci Technol Weld Joining 16(7): 567–574
47. Gaul H, Weber G, Rethmeier M (2011) Influence of HAZ cracks on fatigue resistance of resistance spot welded joints made of advanced high strength steels. Sci Technol Weld Joining 16(5):440–445
48. Wang XP, Zhang YQ, Ju JB et al (2016) Characteristics of welding crack defects and failure mode in resistance spot welding of DP780 steel. J Iron Steel Res Int 23(10):1104–1110
49. Jia S, Zhang Y, Liu X et al (2015) Hot dip galvanized TRIP steel spot welding crack analysis. Electr Weld Mach 45(8):145–149
50. Wang X, Zhang Y, Ju J et al (2016) Effect of resistance spot welding process on welding spot crack defects of advanced high strength steel. Electr Weld Mach 46(6):96–100
51. Yan B, Zhu H, Lalam SH et al (2004) Spot weld fatigue of dual phase steels. SAE technical paper No. 2004-01-0511
52. Kim YG, Kim IJ, Kim JS et al (2014) Evaluation of surface crack in resistance spot welds of Zn-coated steel. Mater Trans 55(1):171–175
53. Barthelmie J, Schram A, Wesling V (2016) Liquid metal embrittlement in resistance spot welding and hot tensile tests of surface-refined TWIP steels. IOP Conference Series-Materials Science and Engineering, vol 118. IOP Publishing, Bristol, pp 1–8
54. Tolf E, Hedegård J, Melander A (2013) Surface breaking cracks in resistance spot welds of dual phase steels with electrogalvanised and hot dip zinc coating. Sci Technol Weld Joining 18(1):25–31
55. Ashiri R, Haque MA, Ji CW et al (2015) Supercritical area and critical nugget diameter for liquid metal embrittlement of Zn-coated twining induced plasticity steels. Scripta Mater 109:6–10
56. Ashiri R, Shamanian M, Salimijazi HR et al (2016) Liquid metal embrittlement-free welds of Zn-coated twinning induced plasticity steels. Scripta Mater 114:41–47

57. Bruscato RM (1992) Liquid metal embrittlement of austenitic stainless steel when welded to galvanized steel. Welding Journal 71:455s–459s
58. Mori H, Nishimoto K (2012) Effect of chromium and nickel contents on liquid zinc embrittlement in heat affected zone of austenitic steels. Q J Jpn Weld Soc 30(1):42–49
59. Pańcikiewicz K, Tuz L, Zielińska-Lipiec A (2014) Zinc contamination cracking in stainless steel after welding. Eng Fail Anal 39:149–154

Part II
Research Papers

Data-Driven Welding Expert System Structure Based on Internet of Things

Chao Chen, Na Lv and Shanben Chen

Abstract With the development of the information technology, the new techniques such as Internet of Things (IOT) and artificial intelligence are introduced into welding manufacturing. This paper introduces a new technique, data-driven welding expert system based on IOT. During welding process, the various sensor information including optical information, electrical information and sound information can be detected to assist welding monitoring. The application of IOT will make it easier to collect and integrate welding information. The data-driven welding expert system can learn and summarize the expert knowledge from these raw welding data without interacting with welding experts. In the end, the paper introduces a structure of data-driven welding expert system based on IOT and demonstrate its function.

Keywords Data-driven · Welding expert system · Multi-sensor information
Internet of things

1 Introduction

It's obvious that, with the fast development of science technology, many new technologies have gradually influenced and changed industrial manufacturing. Nowadays, the fast development of industrial manufacture is based on the development of nine technologies [1], which are big data and analytics [2, 3], autonomous robots [4], simulation [5], horizontal and vertical system integration [6], the internet of things (IOT) [7–9], cybersecurity [10], the cloud, additive manufacturing [11] and augmented reality [12]. Because of the development of these emerging technologies, great changes have come into the welding manufacturing industry. This paper introduces a new technique, data-driven welding expert system based on

C. Chen · N. Lv (✉) · S. Chen
Intelligentized Robotic Welding Technology Laboratory, School of Materials Science
and Engineering, Shanghai Jiao Tong University, Shanghai 200240, China
e-mail: nana414526@163.com

© Springer Nature Singapore Pte Ltd. 2018
S. Chen et al. (eds.), *Transactions on Intelligent Welding Manufacturing*,
Transactions on Intelligent Welding Manufacturing,
https://doi.org/10.1007/978-981-10-8330-3_3

IOT, which will be demonstrated in the follow three sections. Section 2 introduces the application of IOT in welding manufacture industry. Section 3 demonstrates the various welding information sensor techniques that includes electrical information, optical information and sound information. Section 4 introduces the application of expert system in welding manufacture and mainly elaborates the data-driven welding expert system based on IOT. Section 5 concludes the whole paper.

2 Welding Manufacture Based on IOT

With the technique development of Internet of Things (IOT), the production mode of the welding manufacturing industry has changed a lot. Through doing a series of researches about how IOT affects manufacturing industry, we learned that Shi Yong Wang use a brief framework of the smart factory based on IOT [13]. In the framework, the smart factory consists of four layers: physical resource layer, industrial network layer, cloud layer, and supervision and control terminal layer. As shown in Fig. 1, physical resource layer is based on smart device that can communicate with each other through industrial network. Various information systems, such as manufacturing execution system (MES) and enterprise resource planning (ERP), exist in the cloud that can acquire massive data from the physical resource layer and interact with people through the terminals. This actually forms a cyber-physical systems (CPS) where physical artifacts and informational entities are deeply integrated.

Zuehlke [14] uses a pyramid framework to demonstrate the architecture of intelligent factory based on IOT, from field devices (sensors/actuators) and programmable logic controllers (PLC) through process management and manufacturing execution systems (MES) to the enterprise level (ERP) software. As is shown in Fig. 2.

Kemppi's ArcQ [15] all-round welding quality management system is the first integrated system that systematically integrated the welding manufacture and IOT. Its implementation figure is as Fig. 3. ArcQ's management of welding quality is reflected in the following areas:

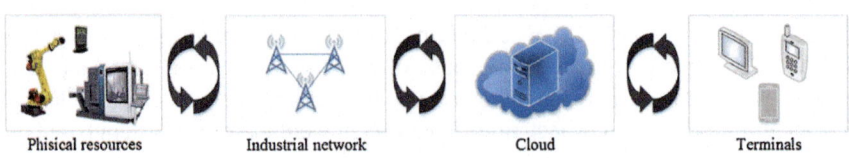

Phisical resources   Industrial network   Cloud   Terminals

Fig. 1 A brief framework of the smart factory of IOT

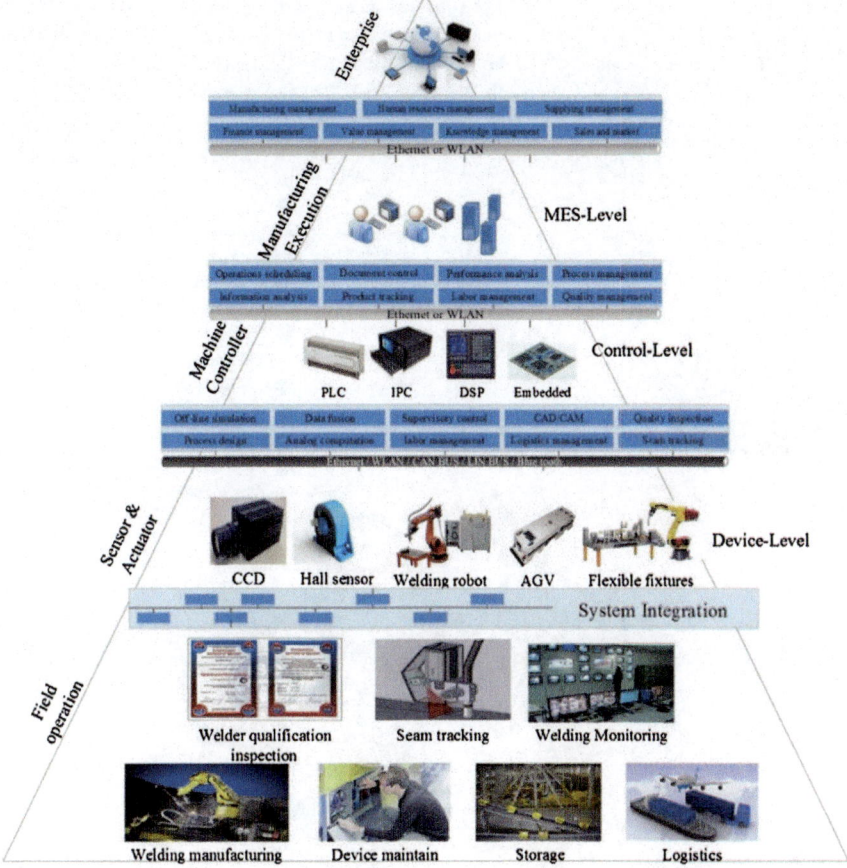

Fig. 2 A pyramid framework architecture of the intelligent factory base on IOT

1. A comprehensive and in-depth management to welding operators. ArcQ identify the qualification of welder at the beginning of welding to assure the welding is completed by correct, qualified and capable welder.
2. Based on WPS, keeping all welding operation following WPS standard. All ArcQ parameters come from WPS. All welding data will be compared to WPS demanding data.
3. Strict management to the use of welding materials. ArcQ demands that welding can start only after check welding stick, welding materials and process number. ISO3834 demands traceability to used materials.
4. Implementing all-time record and management to welding machine's pre-maintenance, use condition, use time. ArcQ can provide welding machine's all pre-use and after-use report to users and inspection department.

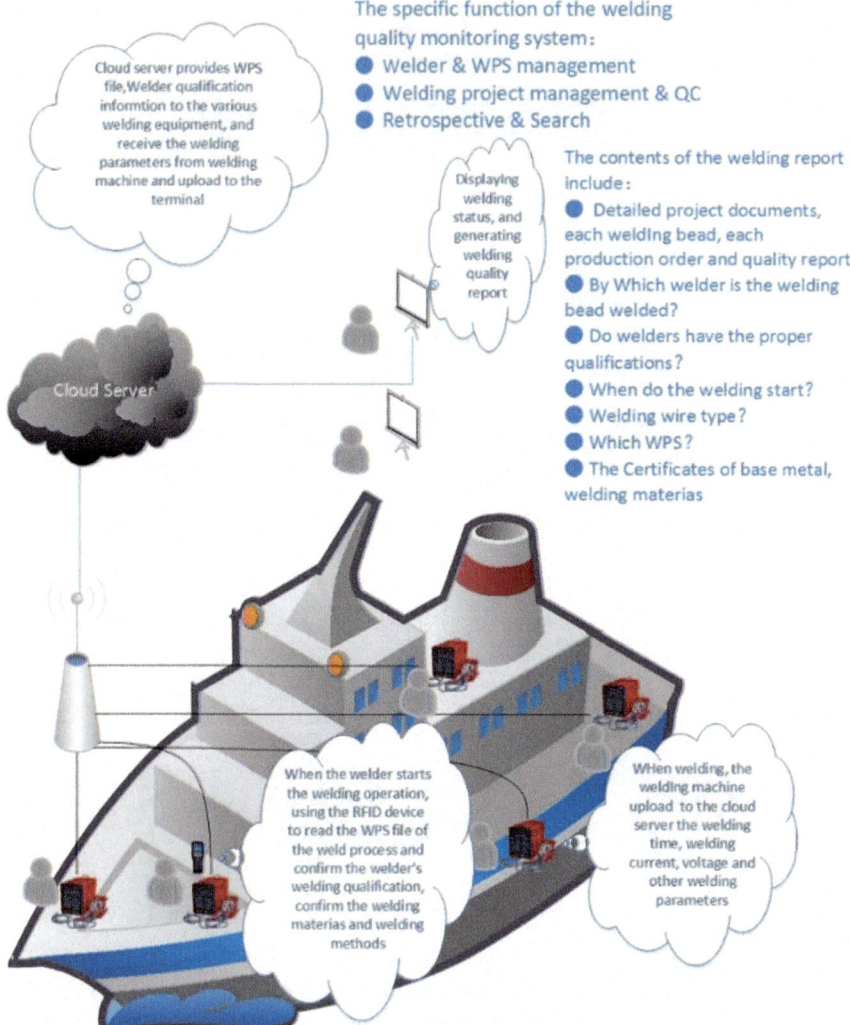

The specific function of the welding quality monitoring system:
- Welder & WPS management
- Welding project management & QC
- Retrospective & Search

The contents of the welding report include:
- Detailed project documents, each welding bead, each production order and quality report
- By Which welder is the welding bead welded?
- Do welders have the proper qualifications?
- When do the welding start?
- Welding wire type?
- Which WPS?
- The Certificates of base metal, welding materias

Cloud server provides WPS file, Welder qualification informtion to the various welding equipment, and receive the welding parameters from welding machine and upload to the terminal

Displaying welding status, and generating welding quality report

Cloud Server

When the welder starts the welding operation, using the RFID device to read the WPS file of the weld process and confirm the welder's welding qualification, confirm the welding materias and welding methods

WHen welding, the welding machine upload to the cloud server the welding time, welding current, voltage and other welding parameters

Fig. 3 ArcQ all—round welding quality management system

5. Full-process and full-time record and real-time delivery of all demanding welding parameters. Through above feedback, ArcQ can shorten time that problem is found and reduce the loss caused by problem. Not only can welding quality be improved, but also efficiency can be enhanced.

6. Only after dealing with all welding deviation, welding can proceed. When the welding deviation occurs, the system promptly alarm, ArcQ requires that the qualified operators must deal with the deviation, or the error record will remain unsettled state. ISO3834 has the same strict demand too.

7. Strict limit to operating authorization of welding operators, managers, technicists. Only by doing this, the dealing of welding deviation can be completed by correct operators, which conforms to ISO3834 demands.
8. Complement quality report. All WPS required parameters during welding can be recorded in units of 200 ms and can be permanent preservation.
9. Offering qualification management and performance appraisal for welding managers and human resources. Welding workpiece number, welding success rate, deviation number, deviation causes. These data can be used in manufacturing management and individual performance management.

3 Welding Information Sensing and Analyzing

Due to the development of IOT, the welding process data can be obtained and integrated more easily. There are a lot of researches about welding information sensing technique. In [16, 17], Shanben Chen presents a systematic overview on multi-information sensing technique of arc welding dynamic process. In [18], Zhifen Zhang demonstrates multisensory data fusion technique and its application to welding process monitoring. A series of welding information sensing technique are discussed as follows:

3.1 Electrical Information

As far as the electrical information is concerned, welding current and arc voltage is the parameters detected most commonly. As both of them have good ability of detecting the abnormal welding state such as excess or lack of welding gas, mismatched welding feed rate and so on [19]. Despite all this, it is still difficult to distinguish the specific factors that influence the normal welding state only by detecting and analyzing the welding current and arc voltage [20, 21].

3.2 Optical Information

The type of welding optical information includes the optical spectrum emission and vision information. The two types of optical information will be demonstrated from academic perspectives as follows.

Optical Spectrum Emission
Optical spectrum emission information has a lot of advantages. First, it contains abundant welding process information, which include spectral lines of the various particle in arc plasma and black-body radiation spectrum line of electrodes, molten

metal and protective gases. Second, optical spectrum emission has a high sensitivity and accuracy. Besides, there is no direct influence on the welding system during acquired process of optical spectrum emission information. So the objectivity and accuracy of measured information can be guaranteed. There are two main method that extract the features. The one is extracting the plasma physical characteristic parameters such as arc temperature, electron density by analyzing wavelength and intensity of specific spectrum line. For instance, Mirapeix [22] measured electron temperature of corrosion resistant plate AISI-304 by using relative intensity temperature measurement of two spectral lines during welding process. The other is obtained geometric, morphological and statistical parameters such as peak value, spectral linewidth and intensity mean value through analyzing specific or many spectral line or section of spectral line. For example, Shea [23] realized a real-time Ar arc H concentration detection system by using the intensity comparison between 656.3 nm HI spectral line and 696.5 nm ArI spectral line. Sibillano [24] found that there is a strong relevance between weld seam surface oxidation layer caused by loss of Al and spectral area of Al(II) of 559.79, 625.04, 704.73 nm wave length and Mg(II) of 571.69, 766.9, 789.70 nm wave length, by analyzing active plasma arc spectral line during AA5083 aluminium alloy laser welding.

Vision Information

Welding visual information is mainly from the visible light information, which can directly reflect the dynamic change of welding pool and weld seam. It has abundant information which includes welding pool state, arc form, weld seam position, joint type and so on. The welding visual information can be divided into two categories according to the difference of objective light source. They are active mode and passive mode [25]. The visual sensor can be divided into the two dimensional plane mode and the three dimensional cubic mode according to image feature acquired by visual sensor system [26, 27]. In recent years, the research hotspot mainly focus on the weld seam shaping control [28], weld seam tracking [29], the initial welding position guiding [30] and welding defect detecting [31]. That extracts passive visual information mainly depends on the arc light and welding pool black-body radiation. Xu [32] developed a real-time seam tracking control technology base on passive vision system in robotic gas tungsten arc welding. The active visual information need the extra light source such as laser which is used to illuminate weld seam. And the geometrical information of weld seam and weld pool can be obtained by analyzing reflected laser stripes. Song [33] designed a set of welding pool visual detecting system combined by dot matrix laser and high electronic shutter camera. The clear dot matrix reflected image of welding pool can be obtained through this detecting system. In [34], ShanBen Chen establishes a welding robot system with single camera fixed on the weld torch end-effector for the robot to identify the dimensional position of typical weld seam by one-item and two-position method. It can be used to acquire weld seam dimensional position information in welding robot system.

3.3 Sound Information

The sound signal can be divided into AE(Acoustic emission) and AS(Audible sound). The acoustic emission signal [35, 36] is elastic stress wave signal taking place during the plastic deformation occurs inside of material. Its frequency can reach dozens of million Hz. Kannatey-Asibu [35] applied the acoustic emission sensor technology during the process of arc welding and laser welding to monitor the welding state. The audible sound signal can be transferred into voltage signal through vibrating membrane of the sound sensor. The processing method can be mainly divided into two categories: the one is extracting and analyzing feature in time domain, frequency domain and time-frequency domain; the other is extracting feature through building arc sound channel mathematical model by using LPC method. There is many researches about arc sound during the welding process. Arata et al. [37] tried to extract welding sound feature and explore the effect of welding parameters on welding sound. Lv [38] implemented a real-time monitoring system of welding path in pulse metal-inert gas robotic welding using a dual-microphone array.

3.4 Other Welding Information

Besides the above common welding information, there are many other types of welding information, such as temperature information, ultrasonic information and so on. They reflect dynamic welding process from different perspective, delivering welding quality information directly and indirectly. For example, Nagarajan [39] realized the real-time monitoring of welding dynamic process based on infrared sensing technique. In [40], the designed real-time ultrasonic testing device based on resistance spot welding can detect the broken welding spot and can reach 100% accuracy. However, duo to the complexity of welding process and limitation of sensing medium, every welding sensing technique has its own limitation.

4 Welding Expert System Structure

4.1 Traditional Welding Expert System Structure

The expert system is a computer software system that uses a knowledge repository that solves practical problems that can't be solved in a particular way in a particular domain. It is a branch of artificial intelligence application [41]. It is characterized by the ability that program expert knowledge of various fields. By relative inference method of the expert system program, the ordinary operators can input the initial data and question. Then the expert system can give the operator the relative

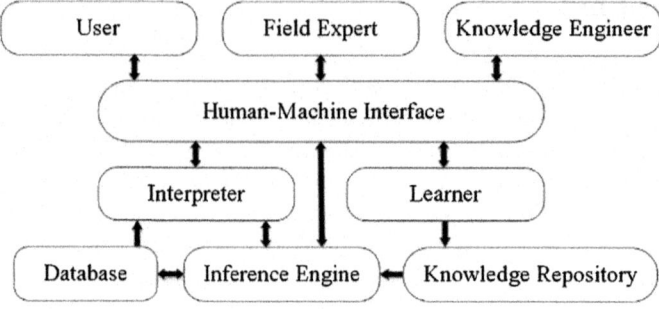

Fig. 4 Traditional welding expert system structure

conclusion of expert level. The traditional expert system structure is shown as Fig. 4. As the welding is a very complex process that need various decision information, which causes that many welding parameters cannot be confirmed through quantitative functional equation unless that giving the quantitative conclusion after qualitative judging welding condition. Nowadays, there are many examples that the expert system is applied to the welding field as shown in Table 1.

By analyzing the various welding expert system, we can find that the welding expert system can be divide into seven types according to the different function:

Table 1 Welding expert system example

Name	Type	Source	Country
Welder qualification	Welder qualification test	Danish Welding Institute	Denmark
Weldcrack expert	Defect diagnose	TWT	Britain
PREHEAT PLUS	Defect diagnose	Edison Welding Institute	America
Weld estimating	Cost estimation	Stone&Webster Engineering	America
Weld procedure selection	Welding process selection	Stone&Webster Engineering	America
Weld assist	Welding process design	Kuhne Cary and Printy	America
Weldex	Welding process design	Technical University of Berlin	Germany
SAW—Ship building environment	Welding process design	Queen University of Belfast	Britain
Weldgen	Welding process design	TW1	Britain
Weld symple	Welding structure CAD	CSM	America

welding process designing type [42], defect diagnosing or equipment failure diagnosing type [43, 44], real-time welding monitoring type [45], welding CAD designing type [46], welder qualification testing type [47], welding robot equipping type [48]. And the ultimate purpose of various welding expert is to control shape and property of welding products.

4.2 Data-Driven Welding Expert System Structure Based on IOT

Due to the powerful welding sensor technology during welding monitor process, on average, about 0.5 GB data will be generated per weld seam. As for the heavy industry and mass production industry, there will dozens of TB data will be generated every day. As the technique development of IOT, the data acquiring process will be easier and the method of data analysis will be more advanced.

As the welding data increases, the welding statistical feature will be more obvious and play a more important role in welding process analysis. However, it is obvious that the knowledge repository of the traditional welding expert system shown in Fig. 4 can't learn from these actual manufacture data and sensor data. It can only update its knowledge repository by interacting with welding expert and welding engineer. The Human-machine Interface module and Learning module of welding expert system can transfer the welding expert knowledge into the form that the computer can understand. The data-driven welding expert system can update its knowledge repository through analyzing and summarizing these welding data.

The Key Technique and Hot Topics
In Fig. 4, it is obvious that the core of traditional welding expert system is its knowledge repository and inference engine. The traditional knowledge repository is used to restore the knowledge offered by experts and engineers. And the knowledge presentation technology of expert system includes rules, semantic net, framework, script and language expressing knowledge such as KL-1, KRYPTON and concept map. In the data-driven welding expert system, the knowledge is created through analyzing welding manufacturing data. The data-driven expert system can: (1) extract effective data from manufacturing noise data; (2) transfer data into information; (3) transfer information into knowledge; (4) summarize the knowledge into the meta-knowledge. So it is one of the key technique to extract meta-knowledge from noise data.

Inference engine can obtain the conclusion aiming at the current problem according to the known information and condition input by users. There are two reasoning methods: forward reasoning and backward reasoning. The strategy of forward reasoning is to find out the rules that can match the input condition and to use the conflict elimination strategy to select one of these satisfied rules to change the contents of the original database. This is done repeatedly, until the database's facts are consistent with the goal, finding the answer, or stopping when there are no rules that match it. The strategy of backward reasoning is to proceed from the selected target and find the rule that the consequence can be achieved. If the premise of this rule matches the fact in the database, the problem is solved; otherwise, the precondition of this rule is regarded as new Sub-goals, and find new sub-goals can be applied to the rules, the implementation of reverse sequence premise, until the last rule of the premise can match the facts in the database, or until no rules can be applied, the system will be dialogue The form asks the user to answer and enter the necessary facts. The design of inference engine is another technique point of data-driven welding expert system.

The Structure Design of Data-driven Welding Expert System

The structure of data-driven welding expert system is shown in Figs. 5 and 6. Two kinds of expert system are exhibited such as welding process designing expert system and welding monitoring expert system. Distinguished from the traditional welding expert system, the data-driven welding expert system can update its Predicted System module (the Interpreter module in Fig. 4) by learning from welding manufacture data. In this structure, the Classification/Regression Model module substitutes the traditional inference engine that can only understand human expert knowledge.

The core of data-driven welding expert system knowledge repository is its data processing method that can analyze the raw welding data and induct the proper expert rules. The machine learning method plays an important role in data-driven welding expert system. Machine learning is a field of computer science that gives computers the ability to learn without being explicitly programmed [49]. Machine learning method can be classified into three categories according to their learning "signal" and "feedback". They are supervised machine learning, unsupervised machine learning, reinforcement machine learning. Supervised learning is the machine learning task of inferring a function from labeled training data [50]. The common approaches and algorithms include artificial neural network, decision tree

Fig. 5 Welding process
designing expert system

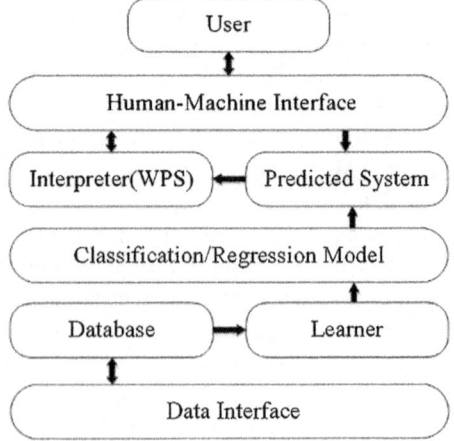

Fig. 6 Welding monitoring
expert system

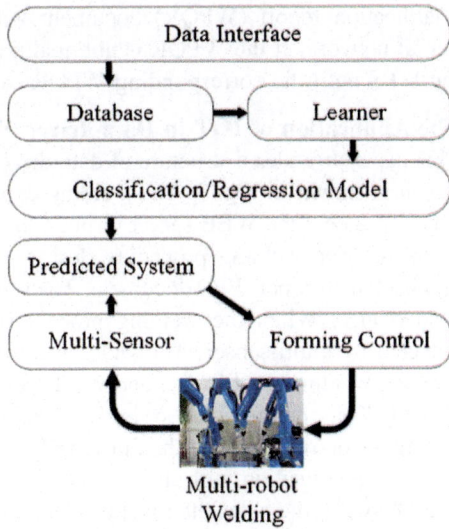

Multi-robot
Welding

learning, support vector machines, random forest and so on. Lv [51] applied a BP-Adaboost Model to predict welding penetration state during pulse GTAW process. Zhang [52] implement a multisensory-based real-time quality monitoring for Al alloy in arc welding by means of SVM-CV wrapper. Unsupervised machine learning is the machine learning task of inferring a function to describe hidden structure from "unlabeled" data (a classification or categorization is not included in the observations) [50]. The common approaches to unsupervised learning include K-means, mixture models, PCA, manifold learning, t-SNE and so on. Wu [53] used t-SNE and DBN model to monitor VPPAW penetration state based on fusion of visual and acoustic signals. Huang [54] used an improved K-medoids algorithm to select the arc spectral line of interest. Reinforcement learning (RL) is an area of machine learning inspired by behaviorist psychology, concerned with how software agents ought to take actions in an environment so as to maximize some notion of cumulative reward.

There are many researches about data-driven welding expert system. Wang [55] designed a GTAW procedure expert system based on neural network. The welding expert system can present the welding procedure specification (WPS). And its database design was based on the C/S mode. The neural network model was established to implement the welding procedure design. The welding expert system can update its function accuracy by expanding its welding case in database. Zhan [56] designed an intelligent welding procedure qualification system for Q345R SMAW. The system consisted of three sub-system, welding procedure design expert system, welding procedure document manage system and prediction system with artificial neural networks. The welding procedure design expert system can generate WPS document according to the user's initial input condition. Then the WPS document was input into the prediction system and the welding procedure qualification report (WPQR) document will generate by calculation of artificial neural network. If the WPQR is not qualified, the system will demand redesigning the WPS until the corresponding WPQR can meet the requirement.

The Application of IOT in Data-driven Welding Expert System

This paper provides a structure of data-driven welding expert system based on IOT which is shown in Fig. 7. The welding expert system is based on B/S mode. The user can access the WEB server through the browser and submit the initial welding demands. The welding procedure design expert system on the WEB server will feedback the proper WPS document. The user can present the WPS document to the field device. When the welding process starts, the welding information can be detected by multi-sensor. The welding information including welding current, arc voltage, welding pool image and weld seam image will be sent into the predicted system. Then the predicted system will give out welding qualification report to the Forming Control module. The Forming Control module will compare the calculated welding qualification with the demands of WPS. By real-time rectifying the deviation between the calculated welding qualification and WPS demands, the welding expert system can obtain the aim of controlling welding formation and property. More than that, the welding expert system can self-learning by updating its database according to actual welding case. The bigger data size the database has, the more accurate the expert system will be.

And the display interface is shown in Fig. 8. On the browser, we can manage relevant welding operators and welding material. The WPS document can be generate according to input weld condition by users and can be appointed to the corresponding welding project. The welding process information and WPQR document of corresponding weld project will be send back and displayed on the browser. These data will be restored in the database and serve as the weld raw data to update the data-driven knowledge repository of welding expert system.

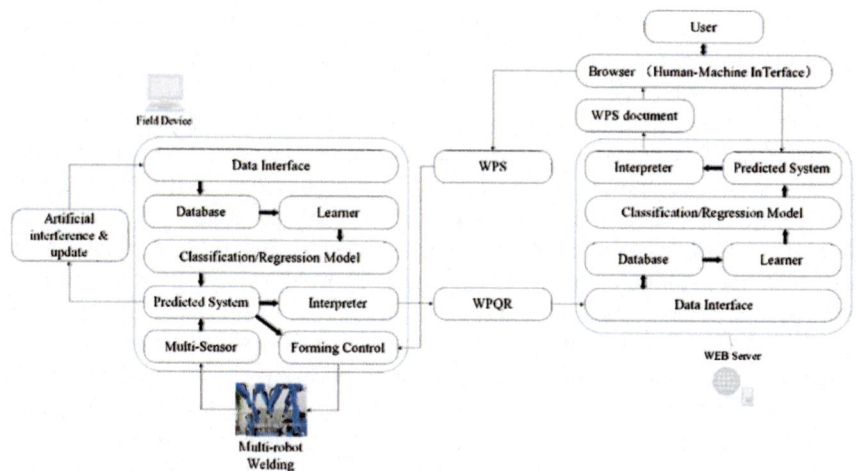

Fig. 7 The structure of data-driven welding expert system based on IOT

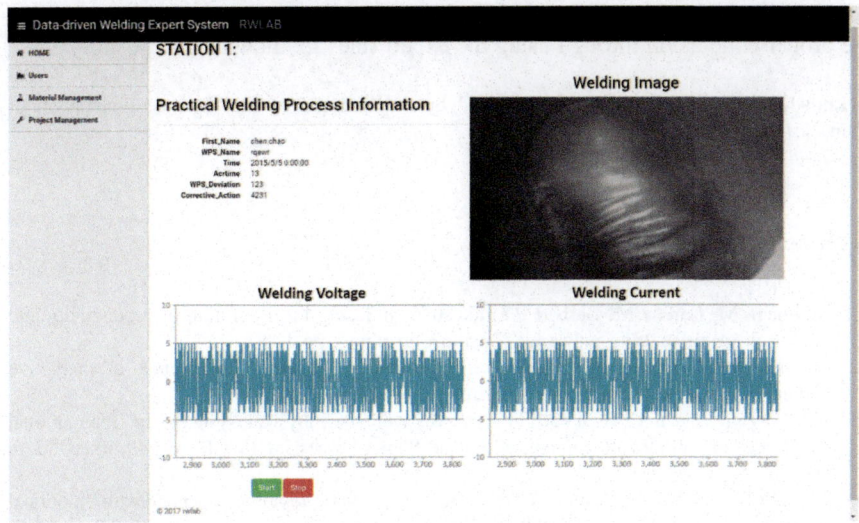

Fig. 8 The display interface of data-driven welding expert system based on IOT

5 Conclusion

With the development of the weld technique, there comes a lot of technique innovation in the welding manufacture industry. The strong sensor technique makes it more convenient to get welding process information including welding current, arc voltage, welding sound and optical information. The innovation of IOT accelerates the process of data collection and integration. As data scale increases, the statistical features of welding data can't meet the requirement of welding manufacture. The artificial intelligence plays a more important role in welding manufacture industry as it can learn from welding data and provide more information and function. As a fatal branch of artificial intelligence, the expert system can assist welding manufacture effectively. The main goal of welding expert system is to control welding formation and property. However, the traditional welding expert system can't update its knowledge repository unless depending on the interaction of filed expert and welding engineer. So the data-driven welding expert system will be a trend in future because it can learn from the raw welding case and summarize its expert knowledge to update its knowledge repository. So the paper constructs a structure of data-driven welding expert system based on IOT.

The core problem of data-driven welding expert system is its establishment of knowledge repository and designation of inference engine. In the future work, we need to combine the expert system and data-driven method such as machine learning effectively. The establishment premise of specific self-learning knowledge repository is to make data-driven method extract the expert rules from welding noise data.

And the matched inference machine mechanism need to be designed to calculate out the proper conclusion through using the expert rules in knowledge repository.

Acknowledgements This work is supported by the National Natural Science Foundation of China (61401275, 61374071 and 51405298).

References

1. Rüßmann M, Lorenz M, Gerbert P et al (2015) Industry 4.0: the future of productivity and growth in manufacturing industries. Boston Consulting, pp 1–5
2. Lee J, Lapira E, Bagheri B et al (2013) Recent advances and trends in predictive manufacturing systems in big data environment. Manuf Letter 38:41
3. Shi J, Wan J, Yan H et al (2011) A survey of cyber-physical systems. In: International conference on Wireless Communications and Signal Processing (WCP), vol 49, issue No. 6, pp 1–6
4. Leconte F et al (2016) Design and integration of a spatio-temporal memory with emotional influences to categorize and recall the experiences of an autonomous mobile robot. Auton Robots 40(5):831–848
5. Barsoum Z, Lundbäck A (2009) Simplified FE welding simulation of fillet welds—3D effects on the formation residual stresses. Eng Fail Anal 16(7):2281–2289
6. Pudjianto D, Ramsay C, Strbac G (2007) Virtual power plant and system integration of distributed energy resources. IET Renew Power Gener 1(1):10–16
7. Tao F, Zuo Y, Xu LD et al (2014) IoT-Based intelligent perception and access of manufacturing resource toward cloud manufacturing. IEEE Trans Industr Inf 10(2):1547–1557
8. Jing Q, Vasilakos AV, Wan J et al (2014) Security of the internet of things: perspectives and challenges. Wirel Netw 20(8):2481–2501
9. Chen F, Deng P, Wan J et al (2015) Data mining for the internet of things: literature review and challenges. Int J Distrib Sens Netw 9:12
10. Ten CW, Manimaran G, Liu CC (2010) Cybersecurity for critical infrastructures: attack and defense modeling. IEEE Trans Syst Man Cybern-Part A: Syst Hum 40(4):853–865
11. Kruth JP, Leu MC, Nakagawa T (1998) Progress in additive manufacturing and rapid prototyping. CIRP Ann-Manuf Technol 47(2):525–540
12. Aiteanu D, Hillers B, Graser A (2013) A step forward in manual welding: demonstration of augmented reality helmet. In: 2013 IEEE International Symposium on Mixed and Augmented Reality (ISMAR), vol 2013. Tokyo, pp 309–310
13. Wang S et al (2016) Implementing smart factory of Industrie 4.0: an outlook. Int J Distrib Sens Netw 12(1):3159805
14. Zuehlke D (2010) SmartFactory—towards a factory-of-things. Annu Rev Control 34(1): 129–138
15. Dupriez Nataliya Deyneka, Truckenbrodt Christian (2016) OCT for efficient high quality laser welding. Laser Technic J 13(3):37–41
16. Chen SB, Wu J (2009) Intelligentized technology for arc welding dynamic process. In: Lecture notes in electrical and engineering, vol 29. Springer, Germany
17. Chen SB, Lv N (2014) Research evolution on intelligentized technologies for arc welding process. J Manuf Process 16:109–122
18. Zhang Z et al (2016) Multisensory data fusion technique and its application to welding process monitoring. In: 2016 IEEE workshop on advanced robotics and its social impacts. Springer, Shanghai, pp 294–298
19. Madigan R (1999) Arc sensing for defects in constant-voltage gas metal arc welding. Weld J 78:322S–328S

20. Koseeyaporn P, Cook GE, Strauss AM (2000) Adaptive voltage control in fusion arc welding. IEEE Trans Ind Appl 36(5):1300–1307
21. Quinn T, Smith C, McCowan C et al (1999) Arc sensing for defects in constant-voltage gas metal arc welding. Weld J 78:322
22. Mirapeix J, Cobo A, Garcia-Allende PB et al (2010) Welding diagnostics based on feature selection and optimization algorithms. Proc SPIE 7726(4):45008–45014
23. Shea JE, Gardner C (1983) Spectroscopic measurement of hydrogen contamination in weld arc plasmas. J Appl Phys 54(9):4928–4938
24. Sibillano T, Ancona A, Berardi V et al (2006) A study of the shielding gas influence on the laser beam welding of AA5083 aluminum alloys by in-process spectroscopic investigation. Opt Lasers Eng 44(10):1039–1051
25. Song H, Zhang Y (2008) Measurement and analysis of three-dimensional specular gas tungsten arc weld pool surface. Weld J 87(4):85
26. Song H, Zhang Y (2007) Image processing for measurement of three-dimensional GTA weld pool surface. Weld J 86(10):323
27. Song HS, Zhang YM (2007) Three-dimensional reconstruction of specular surface for a gas tungsten arc weld pool. Meas Sci Technol 18(12):3751
28. Zhang YM, Kovacevic R, Li L (1996) Adaptive control of full penetration gas tungsten arc welding. IEEE Trans Control Syst Technol 4(4):394–403
29. Xu YL, Zhong JY, Ding MY (2013) The acquisition and processing of real-time information for height tracking of robotic GTAW process by arc sensor. Int J Adv Manuf Technol 65:1031–1043
30. Ye Z, Fang G, Chen SB (2013) A robust algorithm for weld seam extraction based on prior knowledge of weld seam. Sens Rev 33:125–133
31. Kovacevic R, Zhang Y, Li L (1996) Monitoring of weld joint penetrations based on weld pool geometrical appearance. Weld J 75(10):317–329
32. Xu YL, Yu HW, Zhong JY (2012) Real-time seam tracking control technology during welding robot GTAW process based on passive vision sensor. J Mater Process Technol 212:1654–1662
33. Song HS, Zhang YM (2007) Three-dimensional reconstruction of specular surface for a gas tungsten arc weld pool. Meas Sci Technol 18(12):3751
34. Chen SB et al (2005) Acquisition of weld seam dimensional position information for arc welding robot based on vision computing. J Intell Rob Syst 43(1):77–97
35. Kannatey-Asibu E Jr (2009) Principles of laser materials processing. Wiley, Canada, pp 433–434
36. Emel E, Kannatey-Asibu E (1988) Tool failure monitoring in turning by pattern recognition analysis of AE signals. J Manuf Sci Eng 110(2):137–145
37. Arata Y, Inoue K, Futamata M et al (1979) Investigation on welding arc sound (report I)—effect of welding method and welding condition of welding arc sound. Transa JWRI 8(1):25–31
38. Lv N, Fang G, Xu Y et al (2017) Real-time monitoring of welding path in pulse metal-inert gas robotic welding using a dual-microphone array. Int J Adv Manuf Technol 90:2955–2968
39. Nagarajan S, Banerjee P, Chin B (1990) Thermal imaging for weld quality control in arc welding processes. Transp Phenom Mater Process 146:171–178
40. Chertov Karloff A, Perez A et al (2012) In-process ultrasound NDE of resistance spot welds. Insight-Non-Destructive Test Condition Monit 54(5):257–261
41. Liao SH (2005) Expert system methodologies and applications—a decade review from 1995 to 2004. Expert Syst Appl 28:93–103
42. Taylor WA (1986) ES to general arc welding procedures. Metal Constr 7:426–431
43. Lucas W, Brightmore AD (1987) ES for welding engineering. Metal Constr 5:254–260
44. Lucas W (1990) Microcomputers packages and ES for the welding engineers. Weld Metal Frabrication 5:206–212
45. Reeves RE (1988) ES Technology—an avenue to an intelligent weld process control system. Weld J 6:33–41
46. Cary HB (1991) Summary of computer programs for welding engineering. Weld J 1:40–45

47. Wang Z et al (2015). Comparison of welder performance qualification rules between Chinese regulation and ASME BPVS Sec.IX-2015. In: ASME Pressure vessels and piping conference, vol 1B. Codes and Standards, Vancouver, p V01BT01A025
48. Kuhne AH, Cary HB, Prinz FB (1987) An ES for robotic arc welding. Weld J 11:21–25
49. Koza JR et al (1996) Automated design of both the topology and sizing of analog electrical circuits using genetic programming. In: Artificial intelligence in design, vol 96. Springer, Dordrecht, pp 151–170
50. Mohri Mehryar, Rostamizadeh Afshin, Talwalkar Ameet (2012) Foundations of machine learning. MIT Press, Massachusetts, pp 35–36
51. Lv N, Zhong J, Chen H et al (2014) Real-time control of welding penetration during robotic GTAW dynamical process by audio sensing of arc length. Int J Adv Manuf Technol 74(1–4):235–249
52. Zhang Z, Chen S (2017) Real-time seam penetration identification in arc welding based on fusion of sound, voltage and spectrum signals. J Intell Manuf 28(1):207–218
53. Wu D, Huang Y, Chen H et al (2017) VPPAW penetration monitoring based on fusion of visual and acoustic signals using t-SNE and DBN model. Mater Des 123:1–14
54. Huang Y et al (2016) The selection of arc spectral line of interest based on improved K-medoids algorithm. In: 2016 IEEE workshop on advanced robotics and its social impacts. Springer, Shanghai, pp 106–109
55. Wang XW et al (2014) GTAW procedure expert system based on neural network. Appl Mech Mater 455:425–430
56. Zhan X et al (2016) The feasibility of intelligent welding procedure qualification system for Q345R SMAW. Int J Adv Manuf Technol 83(5):765–777

Point Cloud Based Three-Dimensional Reconstruction and Identification of Initial Welding Position

Lunzhao Zhang, Yanling Xu, Shaofeng Du, Wenjun Zhao, Zhen Hou and Shanben Chen

Abstract Initial welding position guidance is necessary for vision-based intelligentized robotic welding. In this paper, we proposed a point cloud based approach to recognize working environment and locate welding initial position using laser stripe sensor. Calibrated laser sensor can achieve high accuracy in transforming from image coordinate system to camera coordinate system and to robot tool coordinate system with hand-eye calibration. Linear feature based image processing algorithm is developed to extract the position of laser stripe center in subpixel-level accuracy; then trajectory-queue based interpolation is implemented to convert down-sampled laser points to robot base coordinate system in real-time scanning. Identification of workpiece is implemented by segmenting workpieces from the point cloud data in the image. Before segmentation, KD-Tree based background model is constructed to filter out background points; then RANSAC fitting procedure rejects outliers and fits the correct workpiece plane model; and the welding initial position can be found along the weld seam which is the intersection of fitted planes. In verification experiment, workpiece planes and welding initial position can be correctly recognized despite the presence of abnormal noises.

Keywords Intelligentized welding · Welding initial position · Laser vision sensor
Welding robot · Point cloud · KD-tree · Segmentation

L. Zhang · Y. Xu (✉) · Z. Hou · S. Chen
School of Materials Science and Engineering, Shanghai Jiao Tong
University, Shanghai, China
e-mail: ylxu@sjtu.edu.cn

S. Chen
Collaborative Innovation Center for Advanced Ship and Deep-Sea
Exploration, Shanghai, China

S. Du · W. Zhao
State Key Laboratory of Smart Manufacturing for Special Vehicles
and Transmission System, Baotou, China

© Springer Nature Singapore Pte Ltd. 2018   61
S. Chen et al. (eds.), *Transactions on Intelligent Welding Manufacturing*,
Transactions on Intelligent Welding Manufacturing,
https://doi.org/10.1007/978-981-10-8330-3_4

1 Introduction

In intelligentized welding applications, autonomous machines such as welding robot are used as executor. It is necessary to equip welding robot with machine vision sensor to handle uncertain and varying environment [1]. Therefore, machine vision techniques are widely deployed in robot welding application, such as guidance, tracking and quality assessment after weldment. Identification of initial position before welding is one scenario where vision sensor can map the welding environment and eliminate the uncertainty of workpiece location. Various machine vision methods can be divided into two main categories: passive vision sensing, which and active vision sensing. Extensive noise such as reflection and make active sensor a good candidate for weld environment. Besides tracking, guiding robot to initial position of weld seam is a vital procedure before weldment. It requires that robot can map the work piece and workspace environment and locate the desired relative to weld torch.

Some early works investigate passive-vision based localization on weld joint and welding initial position. Zhu and Lin used edge detection and template matching to locate weld seam and its intersection with workpiece edge. To accelerate process and boost precision, they adapt two-step match method, first in global and then in local area [2]. Similarly, Chen use "coarse-to-fine" method to find initial point of weld seam. A global curve is fitting to find the weld seam, which can then be narrowed down to a local window where search of intersection between weld seam and workpiece edge is performed. Three-dimensional position is given by dual-cameras measurement [3]. Stereo vision like dual-camera system are widely used in such weld joint localization scenario. Dinham and Fang propose a stereo approach to locate three-dimensional position of weld joints. In their method, background is subtracted by Hough line detection of workpiece edge, then edge detection is used to extract weld seam. Identification of weld joints cam achieve accuracy within ±1 mm in three-dimensional space based on homography matching technique. Experiments in a working space verifies that this method can be implemented in industry [4]. Passive-vision methods are widely deployed in welding application, Xu et al. developed a welding seam tracking system using single passive vision sensor and achieve high tracking accuracy in real-time [5].

Compared to above-mentioned passive vision sensors, laser stripe based active vision sensor performs well in weld tracking applications, because external laser light can still be observed under intensive arc light. This good characteristic attracts many researchers with various investigations. To track weld seam in environment with intensive noise, Xinde Li and Xianghui Li utilize Kalman filter to track laser stripe after image preprocessing. They also build description model for weld joint profile in character strings. Comprehensive experiment in static precision, real-time and dynamic precision shows that their method outperforms others in static, dynamic precision and stability [6]. Ding and Huang utilized principle of triangulation to derive equations that transform image pixels to two spatial directions: X direction which is along the laser stripe and Z direction as dept. To detect weld joint feature, they propose a correlation coefficient based matching method which compare current frame with last frame's joint profile, and maintain a first-in-first-out queue for refreshing seam position during welding. Their matching algorithm enables detections of different groove types and no pre-defined model is needed [7].

Although tracking using laser stripe sensor is relatively mature, there is few explorations on localization of welding initial position based on laser-stripe sensor system, and stereo vision based localization approach requires extra hardware besides laser stripe sensor, which make system more complex. In this paper, three-dimensional reconstruction of working environment and localization of weld joint will be fulfilled using a spatial point cloud approach. After laser calibration and hand-eye calibration, laser-stripe sensor system mounted on robot end effector can scan the working environment, transform laser pixels to spatial space and identify workpiece and initial welding position before welding.

2 System Setup

As shown in Fig. 1, robotic welding vision system composites of a Fanuc M20-iA industrial robot as executor, which is equipped with weld torch as end effector, and other welding equipment like GMAW weld machine. In-house designed laser stripe

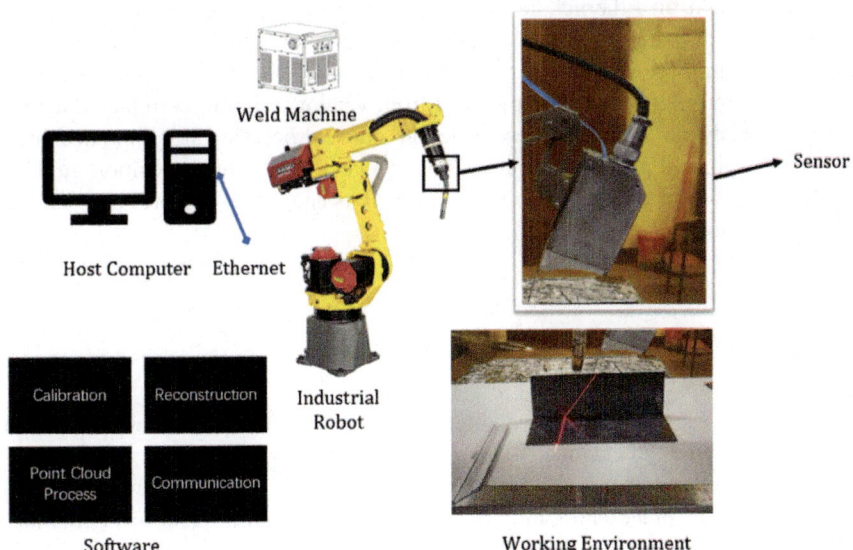

Fig. 1 Laser stripe vision system for reconstruction and localization of initial welding position

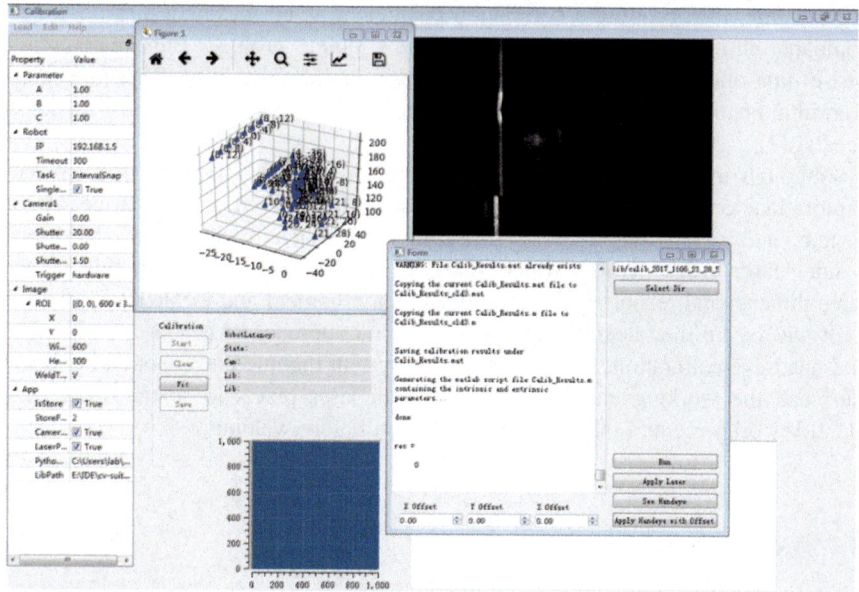

Fig. 2 Graphic user interface of calibration program

sensor is used and mounted in the weld torch. The host computer communicates with robot controller for exchanging robot information and with laser sensor for image data, both on Ethernet. It is the core of the whole system since data like image and robot position are processed by software components running in it.

The software part running on host computer includes communication interface with robot controller based on a protocol from vendor, calibration implementation which is the key part of spatial reconstruction, image processing algorithm designed for laser image, and point cloud processing. Procedure like initial position guiding and tracking is implemented in software, as well as an easy-to-use program for fast calibration, shown in Fig. 2.

3 Calibration

Laser stripe based active sensor can collect image with laser pattern, which can be recovered as three-dimensional information. The accuracy of stripe light sensor depends on calibration procedure. In presented system, it requires two stages of calibration: Laser plane calibration and hand-eye calibration. Both are combined into one by using a planar calibration plate with grid pattern.

Calibrating of laser plane targets at establish transform from two-dimensional image pixel to spatial space represented as a camera based frame. Calibrating the hand-eye matrix is the following steps to derive a matrix as transform from camera frame, as the eye of system, to tool frame, as the hand of system. Tool frame in robotic system is based on weld torch, and its TCP, Tool Center Point, is the weld wire tip. In this paper, these two steps are unified into one calibration procedure, which can be done by a single planar calibration plate. Figure 3 presents overall procedure of calibrating.

To fit laser plane equation in three-dimensional space, at least three non-collinear points are required. These points data can, be achieved using calibration target with specific geometry feature in calibration process, like a calibration target with perpendicular planes, which is difficult to manufacture. In this paper, proposed calibration procedure requires only a simple planar calibration plate which is common in camera calibration. Well-known camera calibration method, proposed by Zhang, is widely adapted in camera calibration. Although laser calibration and hand-eye calibration will not directly depend on result of camera calibration. The external parameter of camera can be used to generate non-colinear data for laser calibration, so that a simple plane can be used as calibration plate, as long as camera take multiple pictures from different view. In summary, points on laser are computed

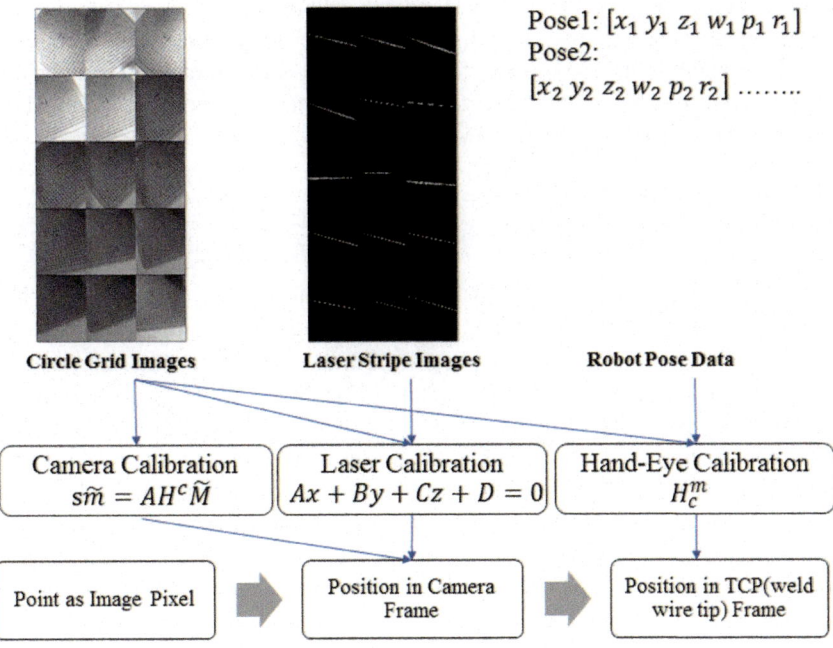

Pose1: $[x_1\ y_1\ z_1\ w_1\ p_1\ r_1]$
Pose2:
$[x_2\ y_2\ z_2\ w_2\ p_2\ r_2]$ ....

Circle Grid Images   **Laser Stripe Images**   **Robot Pose Data**

Camera Calibration
$s\tilde{m} = AH^c\tilde{M}$

Laser Calibration
$Ax + By + Cz + D = 0$

Hand-Eye Calibration
H_c^m

Point as Image Pixel

Position in Camera Frame

Position in TCP(weld wire tip) Frame

Fig. 3 Overall calibration procedure

from intersection of laser line and circle grid pattern. These points from different image at different orientations are all transformed by external parameters into camera frame, and then there are enough non-colinear laser points to fit laser plane equation $Ax + By + Cz + D = 0$. Zhou proposes this method to calibrate laser stripe sensor. Laser plane equation can be combined with camera model to derive relation between two-dimensional laser pixels and three-dimensional position in camera frame. In this paper, a 4 * 3 conversion matrix M_i^c is used to represent this linear transform. The deduction of this conversion matrix is followed Huynh's method, which use homography between laser plane and image plane to derive this linear transform after laser plane is fit [8]. Hand-eye calibration aims at transform from camera frame to tool frame. In this paper, a third-party toolbox is used for computing hand-eye matrix, which is developed by Wengert as a fully automated hand-eye toolbox [9].

Calibration program implement a Graphic User interface which facilitate data acquisition and computing of calibration model. To acquire both laser stripe image and calibration plate image at one pose, program can automatically adjust camera shutter and toggle laser to capture two images once it receives a capture command, one in lower shutter and laser on as laser image, another in higher shutter and laser off as circle grid calibration image. Utilizing the planar calibration target with circle grid pattern, the intersections of laser stripe and extracted circle grid can be calculated and transformed into camera frame. Robot position information is provided by robot controller and recorded for Hand-eye calibration.

In summary, once calibration is finished, the laser pixel in image can be transformed to camera frame, then to tool frame with hand-eye calibration. Then position in robot base frame can be achieved with current robot position. The whole transform can be expressed as follows:

$$P = H^0 H_c^m M_i^c P_i \tag{1}$$

P_i is the homogeneous form of pixel coordinate, like $[x \quad y \quad 1]^T$, M_i^c is the 4*3 conversion matrix from laser calibration, H_c^m is the hand-eye matrix from hand-eye calibration, H^0 is the transform between tool frame and robot base, and usually offered by robot controller as robot position value.

Error both laser calibration and hand-eye calibration. To measure the error, the reprojection method, which use calibrated transform to recalculate the data points, is used. Fig. (4) show the reprojection error of laser plane calibration. For hand-eye calibration, the same reprojection method is used by hand-eye toolbox and result shows that error of approximately 1 mm can be achieved in hand-eye transform, which means hand-eye calibration is the major source of measurement error.

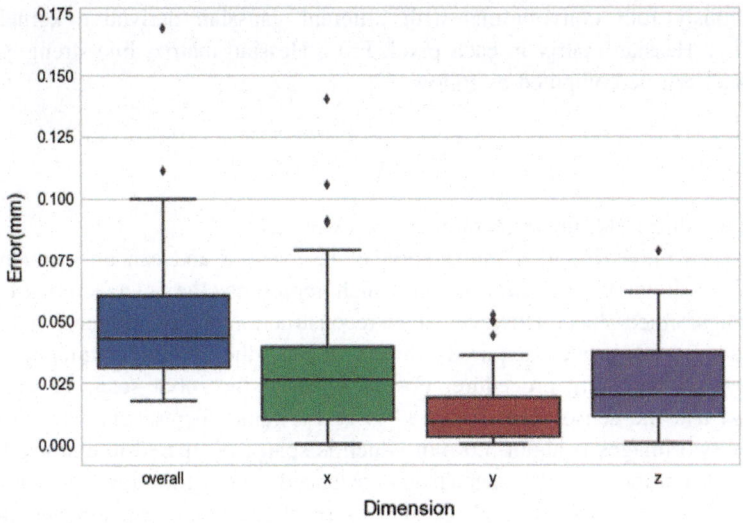

Fig. 4 Reprojection error of laser plane calibration

4 Image Process and Reconstruction

Extraction laser point in two-dimensional image is widely discussed and various methods are proposed. Most of them are based on column-scanning, iterating over the column which is perpendicular to laser line direction and find out the laser point on each column by lightness pattern, like centroid of lightness and gaussian fitting of gray value. Reference [10] proposed a laser stripe peak detector based on FIR filter. Such method's accuracy will easily deteriorate in presence of weld noise, since arc light are far more intense than laser light. Beyond column scanning approach, Steger proposes a robust method to extract linear feature in images, based on Hessian Matrix which describe the property of second-order derivative of image gray values. Du utilize same idea to extract laser stripe in reflective and uneven metallic surface [11]. This hessian-matrix based ridge detect shows good robustness against various noise, and high accuracy at sub-pixel level. This paper will not cover the mathematical intuition behind this method in depth. To extract linear feature, images should be convolved with gaussian kernel, so that $2 * 2$ hessian matrix of each pixel can be achieved. If I is a gray-value image and $r(x; \sigma) = g(x; \sigma) * I$ is convolution operation on I, with $*$ denote convolution operator and $g(x; \sigma) = \frac{1}{\sqrt{2\pi}\sigma} \exp^{\frac{-x^2}{2\sigma^2}}$ as Gaussian kernel. Hessian matrix can be represented by:

$$H = \begin{bmatrix} r_{xx}(x; \sigma) & r_{xy}(x; \sigma) \\ r_{xy}(x; \sigma) & r_{yy}(x; \sigma) \end{bmatrix} \qquad (2)$$

Obviously four convolutions with different gaussian derivative kernel can compute a Hessian matrix in each pixel. From Hessian matrix, line strength N of each pixel can be computed as follow:

$$N = \frac{\sigma^{\gamma}}{2} \left| r_{xx} + r_{yy} - \sqrt{(r_{xx} - r_{yy})^2 + 4r_{xy}^2} \right| \tag{3}$$

Using a threshold, linear feature can be extracted.

Extracted pixels has rich information of laser stripe and can be recovered to three-dimensional profile information which represents the intersection of laser plane and scanned object. However, since reconstruction in three-dimensional space requires a set of images acquired while scanning, the capture frequency is the bottleneck of scanning procedure. For example, if the robot scans at speed of 50 mm/s and the sensor capture speed is at 14 frames per second, the distance between two images is about 3.6 mm which is sparse compared to extracted continuous laser stripe on each image. This unbalanced data distribution can be utilized to reduce its size, which means discarding some information on each image without loss of too many details. Based on this intuition, Down-sample technique is proposed, which will sample laser pixels on each image per M pixels after extraction of laser center pixels. Fig. (5) shows the result of laser extraction without down-sample ($M = 1$), and the result after down-sample with $M = 5$. The laser stripe pixels after down-sample can still retain most of details.

M is the only parameter of down-sample parameter and decides the final data volume. Appropriate M can reduce volume of data with little loss of surface information. In experiment of workspace scanning, M at 20 can still retain most information of workpieces but the total data size is rapidly reduced.

While scanning, robot with sensor will capture a series of images which are discrete both in time and space. To reconstruct these images into three-dimensional robot base frame, it is important to get the precise position where each image captured while scanning, so that we can transform the data from single image to robot base frame. Acquisition of precise robot position where each image captured cannot be done directly since we can only achieve robot position data from controller at a fixed finite frequency, 50 Hz for example. The solution is to maintain a trajectory queue which can linearly interpolate position by time once an image is acquired at certain timepoint.

Fig. 5 Extracted laser center in different M

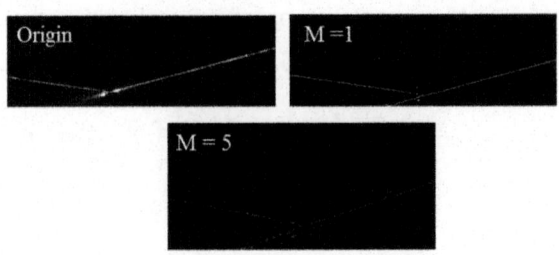

As input of reconstruction, images and robot position are separately handled by host computer but they will both be recorded using a millisecond-level timestamp once program successfully receive these data. However, timestamp like this is the timepoint when communication is finished, not the timepoint when data is generated. For image data from sensor, communication cost is non-trivial and must be considered. This delay can be estimated by image size and ethernet bandwidth. For system in this paper, the image size is 1200 * 1600 * 8bit, and the ethernet connection between host computer and sensor has a bandwidth of 50 Mbit/s, so the communication cost, the timestamp should minus, is about 27.32 ms. As for robot position data streaming, due to its relatively high frequency and unknown internal mechanism, its communication delay is not considered.

For robot position (x_n, y_n, z_n) at timepoint t_n, it is inserted into a trajectory queue after received. This trajectory queue is ordered by timepoint and contains robot position in a fixed duration, such as 500 ms. The purpose of maintaining such a queue is to estimate robot position of arbitrary timepoint.

As shown in Fig. (6), when an image is received and is to be reconstructed in robot base frame, robot position (X_n, Y_n, Z_n) where this image is captured remains unknown. To estimate this position, the timepoint T_n when image is captured is used to interpolate an estimated (X_n, Y_n, Z_n), in following steps:

1. In trajectory queue, use binary search to find the first timepoint t_i which is larger than T_n, now T_n is between t_{i-1} and t_i.
2. Calculate the averaged velocity V_i between t_{i-1} and t_i:

$$\Delta t = (t_i - t_{i-1})$$

$$\begin{cases} V_{xi} = \frac{(X_i - X_{i-1})}{\Delta t} \\ V_{yi} = \frac{(Y_i - Y_{i-1})}{\Delta t} \\ V_{zi} = \frac{(Z_i - Z_{i-1})}{\Delta t} \end{cases} \tag{4}$$

Fig. 6 Interpolation using trajectory queue

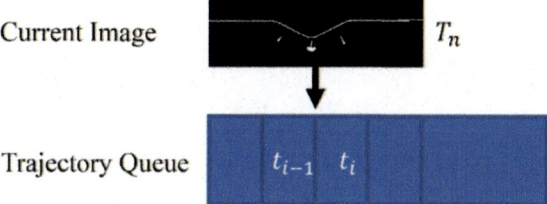

Fig. 7 Reconstruction result
with different scanning speed

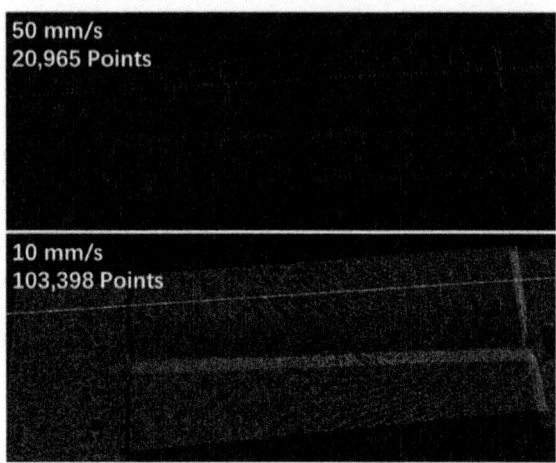

3. Linearly interpolate corresponding position at T_n:

$$\begin{cases} X_n = V_{xi} * (T_n - t_i) + x_i \\ Y_n = V_{yi} * (T_n - t_i) + y_i \\ Z_n = V_{zi} * (T_n - t_i) + z_i \end{cases} \tag{5}$$

Such interpolation based position estimation can recovery robot position in arbitrary timepoint. Therefore, once the timepoint when image is captured is known, the robot position of this image timepoint can be calculated based on the above method, and then the reconstruction in robot base frame, which depends on robot position, can be done by recovery accurate spatial position of a series images captured along scanning direction. This trajectory-queue based interpolation can handle online scanning and reconstruction at various scanning speed. Figure 7 shows reconstructed point cloud after down sampled with $M = 10$ and scanning speed = 50 and 10 mm/s respectively. The green dot is the estimated position where camera captures each image. It is obvious that scanning at 50 mm/s constructs a sparser point cloud, because the camera's capture speed is fixed. Thanks to trajectory queue based interpolation, varying speed will not affect the reconstruction accuracy.

5 Point Cloud Processing

Based on above-mentioned components. Industrial robot mounted with laser sensor can scan welding workspace and acquire three-dimensional point cloud data set in robot coordinate. To further process these points, we utilize PCL (Point Cloud Library), a C++ library for point processing algorithms, to process generated point cloud data.

Robot must move to search before localizing the workpieces. To prevent potential collision between weld torch and work pieces while moving, robot should conform pre-determined search policy. Here is the proposed search scheme: the robot program requires two pre-taught position as its begin and end of search. The search area is defined by the distance between these two points and the field of view of laser sensor, which is affected by laser stripe length and field of view of camera.

In search area, points can be divided into three categories: background, target workpiece and abnormal noise, as shown In Fig. 8. Only workpiece is the interest of later process. Therefore, background and abnormal noise should be eliminated to improve accuracy of identification. Background is what keep static in space, so it will not change in multiple scanning over time. Filtering out these background points can utilize this static feature to construct a background model.

A background model defines geometry of background, and can be used to check whether one point belong to background environment or not. The general ideal is to search every point's nearest k neighbors in background and use distance threshold to reject points that has a high possibility to belong to background, this can be named as kNN distance criterion. To accelerate query of kNN, background model will be constructed as KD-Tree, a data structure offering fast kNN query operation.

KD-Tree is a binary tree data structure, with nodes specifies an axis and splits the set of points based on comparison of their coordinate along and certain value. Once constructed, KD-Tree can be used to queried one point's nearest K points at a time cost of $O(\log N)$, with N for number of points inside tree, which indicates a substantial gain in efficiency. Utilizing kNN (k Nearest Neighbor) query, point can be verified whether it belongs to a background model, by checking its' kth neighbors inside KD-Tree. This distance method is named as kNN Distance method. With a constructed KD-Tree background model, every point in point cloud is checked with kNN Distance criterion. If one point's kNN Distance is smaller than threshold, it will be subtracted from point cloud. Details of kNN Distance are shown in Fig. 9.

Fig. 8 Search area and working environment

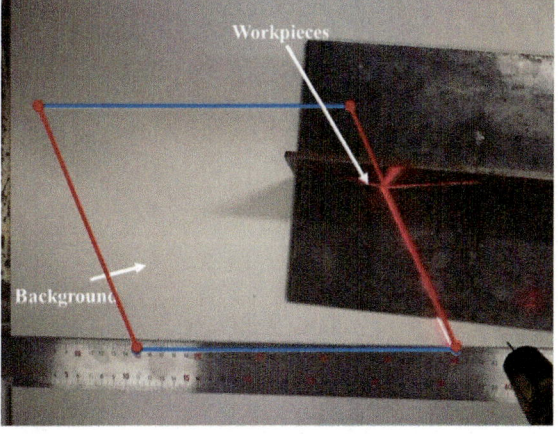

Fig. 9 Filtering based on
kNN distance threshold

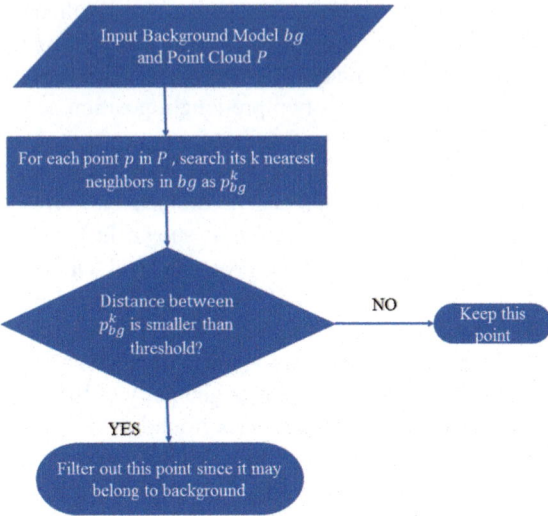

Figure 10 shows effect of background subtraction. The background in working environment is a coarse planar surface and has been constructed before workpiece is put into environment. Then every scanned point cloudas Fig. 10a shown, will use this KD-Tree as input to query every point's kNN Distance in this point cloud. Points in purple in Fig. 10b represents points that have a smaller kNN Distance than threshold so they are marked as background point and will be filtered before later fitting process.

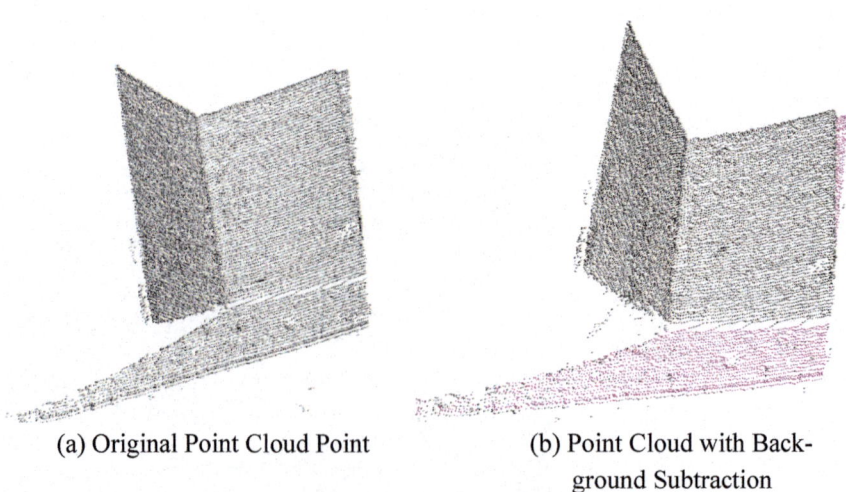

(a) Original Point Cloud Point   (b) Point Cloud with Background Subtraction

Fig. 10 Effect of background substraction

Introduction of KD-Tree based background model and kNN Distance criteria can finish effective background subtraction, eliminate noise for later process.

To locate welding initial position from large point cloud data set, algorithms should be designed to fit different workpieces with different joint profile. The process that separate points belong to workpieces from other points is called segmentation. Segmentation can be based on workpieces' geometry feature. In 3d estimation problem, planar model is a good pattern to fit. Fortunately, most work pieces in welding production and can be describe by composition of different planar model. Therefore, identification of workpieces can be described as a problem of fitting multiple plane model from point cloud data. In this paper, Tee joint with workpieces perpendicular to each other is chosen to as target of recognition.

Although background pints have been subtracted, there is still much noise in point cloud data. To fit a model from noise-polluted data set, a robust fitting algorithm is needed. In this paper, RANSAC (Ransom Sample Consensus) algorithm is used to fit planar workpiece out of point cloud. RANSAC is an iterating method which samples minimum data points from data set to fit model in each iteration, such as three points for plane model. Iteration will terminate if most of data points fit in this sampled model, and outliers can be rejected since they are far from true model. For point cloud data set, noise as outliers will mislead common fitting algorithm such as least-square fitting. Rejecting outliers using RANSAC can improve fitting accuracy greatly by rejecting these outlies over iteration. Fitted plane model can be represented as $Ax + By + Cz + D = 0$, with $\begin{bmatrix} A & B & C \end{bmatrix}^T$ as normal vector of plane.

Tee joint is composed of two workpieces which can be abstracted geometrically as perpendicular planes, and the perpendicular relation is a valuable constraint for fitting procedure: once we find one plane, the other can be easily found by explicitly choose a plane whose normal vector is perpendicular to the previous. This constraint can be expressed as angle θ, which is 90° in perpendicular Tee joint, as the ideal angle relation between two planes, and angle ε as the maximum allowed deviation. Real angle of planes can be computed as angle between two normal vectors of two planes. Real angle and ideal angle can determine whether these planes form a correct weld joint. Criterion is as follows:

$$\left| \cos^{-1} \frac{n_1 * n_2}{n_1 n_2} - \theta \right| < \varepsilon \tag{6}$$

In summary, fitting procedure of Tee joint includes:

1. RANSAC fitting for the first plane P_1 with normal n_1. Subtract all points belong to P_1
2. While fitting the second plane P_2 in RANSAC iteration, reject all fitted planes not satisfied with Eq. (6) using normal vector n_2. Subtract all points belong to P_2
3. For points belong to P_1 and P_2, apply static noise approval to further filter out outliers, then two points clouds represent two planes are achieved.

Fig. 11 Fitting planes of tee joint

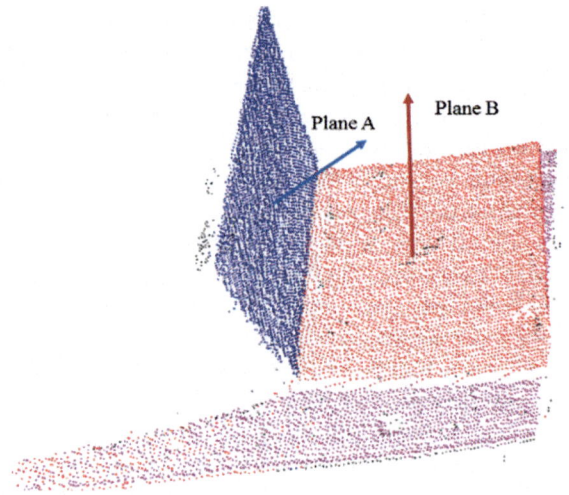

Figure 11 shows the result of Tee joint fitting. Now since planes are segmented from original point cloud, it is simpler to find out weld seam and weld initial position.

The weld seam can be computed as intersection of two plane once planes' coefficients are estimated. Another interested feature, initial welding position, can be figured out along the weld seam. This is a one-dimensional search along intersection. To suppress random noise, compute the centroid of the first x valid point along the seam as the initial position of weld seam. Green line in Fig. (12) represents the weld seam, which is the intersection of two workpieces as Tee joint, and the green dot is the desired initial welding position.

Fig. 12 Identification of weld seam and weld initial position

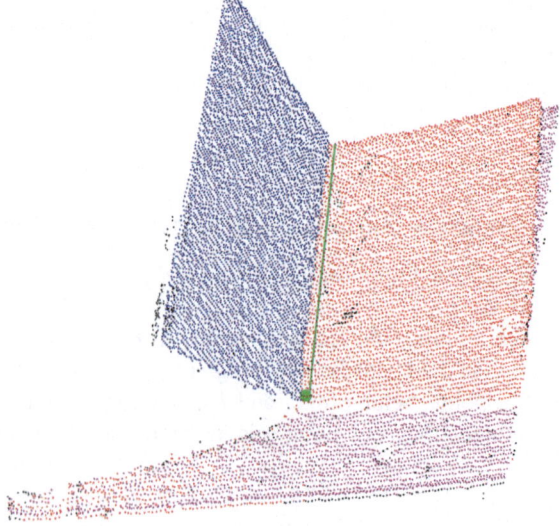

Fig. 13 Working
environment with abnormal
object as noise

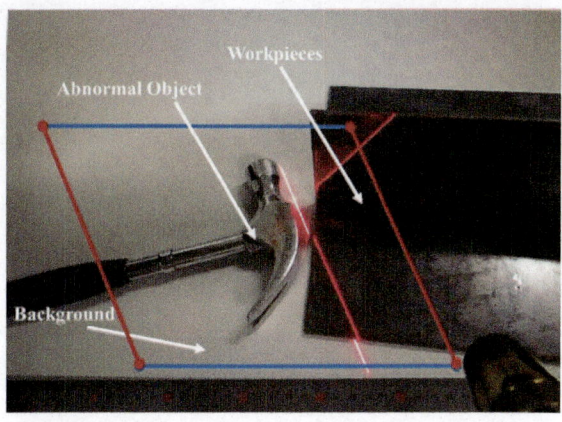

6 Experimental Verification

To verify performance and robustness of proposed algorithm in environment with
unexpected noise, experiment is performed to simulate scenario where both
workpiece and irrelevant object exists. As shown in Fig. 13, a working space with
abnormal object is set up, composed of background, abnormal object and workpieces. As abnormal object, tool is arbitrarily put around workpiece. To search such
a working space, search length is set to 150 mm and search speed is at 50 mm/s.

Before searching, background without other objects is scanned as background
scanning, and KD-Tree model is constructed for kNN Distance based background
filtering, with $K = 3$ and distance threshold of 1.5 mm. In reconstruction procedure,
the down sample is performed at $M = 15$. Recognition result are shown in Fig. 14.
From original point cloud to weld seam and welding initial position, algorithm
presented in this paper can robustly filter out background points and abnormal
object, which shows good practical performance in production environment.

7 Conclusion

In this paper, welding workspace reconstruction and identification of initial welding
position using laser stripe sensor are studied. A robotic welding vision guiding
system is proposed and be tested in noisy working environment. Related works
including:

1. Fast and unified calibration procedure for laser plane and hand-eye conversion is
   proposed and programed as utility software. Laser plane calibration can achieve
   high reconstruction accuracy with approximately 0.1 mm reprojection error, and
   error in hand-eye matrix conversion is about 1.0 mm.

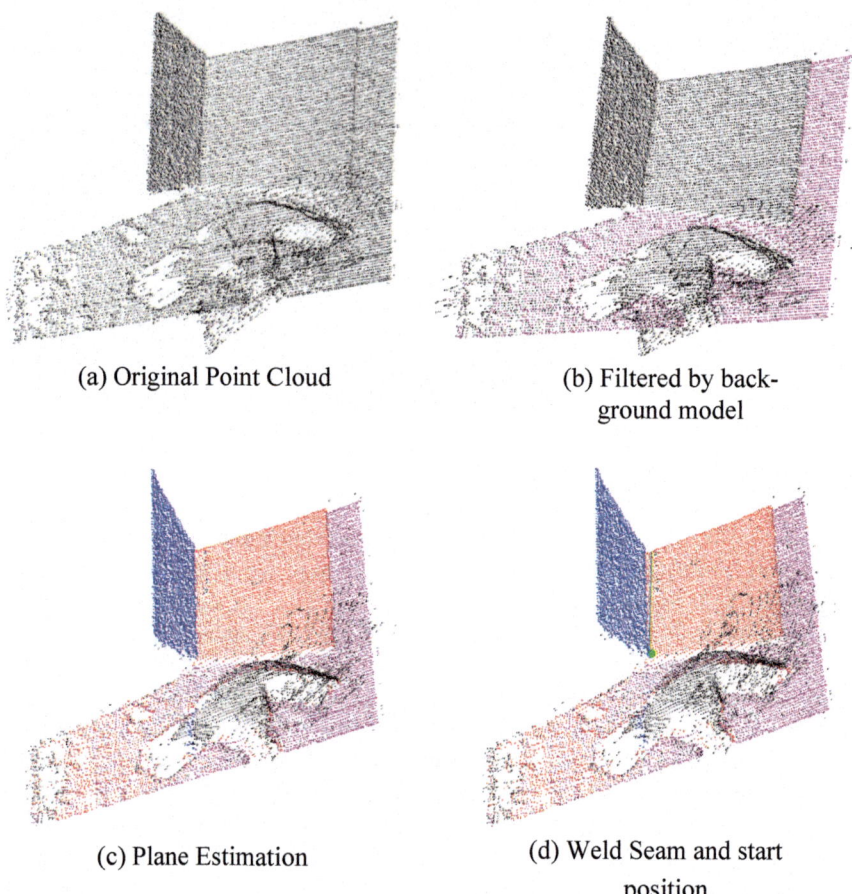

(a) Original Point Cloud

(b) Filtered by back-
ground model

(c) Plane Estimation

(d) Weld Seam and start
position

Fig. 14 Fitting and identification in noisy environment

2. Extracting algorithm for laser stripe center is developed to extract linear feature
   from images with subpixel-level accuracy. Hessian matrix based line extractor
   have good performance in noisy metallic surface.
3. Online reconstruction scheme suitable for moving robot arm is designed with
   two steps: down sample on single image and interpolation based estimation for
   position where image is captured. Three-dimensional Reconstruction from a
   series of images are finished online at different scanning speed.
4. Point cloud based workpiece segmentation is implemented. To segment workpieces out of point cloud, KD-Tree background model and k Nearest Neighbor
   Distance criterion are used to filter out background points, and RANSAC based
   plane fitting approach can fit planes and reject noise. Certain angle relation
   between workpieces is utilized to multiple planes fitting. After workpiece segmentation, initial position can be found along the weld seam, which is the
   intersection of fitted planes.

5. Experimental verification shows system can recognize work pieces and initial position against noise like abnormal object. Robustness of the whole system is proved, and system can be deployed in industrial production environment.

Acknowledgements This work is partly supported by the National Natural Science Foundation of China (51405298 and 51575349), Collaborative Innovation Center for Advanced Ship and Deep-Sea Exploration, the State Key Laboratory of Smart Manufacturing for Special Vehicles and Transmission System (GZ2016KF002), the Development Fund Project of the Science and Technology Committee of Qingpu District and the National High Technology Research and Development Program 863 (2015AA043102).

References

1. Chen SB (2015) On intelligentized welding manufacturing. In: Robotic welding, intelligence and automation. Springer, Berlin, pp 3–34
2. Zhu ZY, Lin T, Piao YJ et al (2005) Recognition of the initial position of weld based on the image pattern match technology for welding robot. Int J Adv Manuf Technol 26(7–8): 784–788
3. Chen XZ, Chen SB (2010) The autonomous detection and guiding of start welding position for arc welding robot. Ind Robot 37(1):70–78
4. Dinham M, Fang G (2013) Autonomous weld seam identification and localisation using eye-in-hand stereo vision for robotic arc welding. Robot Comput Integr Manuf 29:288–301
5. Xu Y, Yu H, Zhong J et al (2012) Real-time seam tracking control technology during welding robot GTAW process based on passive vision sensor. J Mater Process Technol 212(8): 1654–1662
6. Li X, Ge S, Khyam MO et al (2017) Automatic welding seam tracking and identification. IEEE Trans Ind Electron 99:1
7. Ding Y, Huang W, Kovacevic R (2016) An on-line shape-matching weld seam tracking system. Robot Comput Integ Manuf 42:103–112
8. Huynh DQ, Owens RA, Hartmann P (1999) Calibrating a structured light stripe system: a novel approach. Int J Comput Vision 33(1):73–86
9. Wengert C, Reeff M, Cattin PC, et al (2006) Fully automatic endoscope calibration for intraoperative use. Bildverarbeitung für die Medizin, 419–423
10. Forest J, Salvi J, Cabruja E, et al (2004) Laser stripe peak detector for 3D scanners. A FIR filter approach. In: Proceedings of the 17th international conference on pattern recognition, IEEE, pp. 646–649
11. Du J, Xiong W, Chen W, et al (2015) Robust laser stripe extraction using ridge segmentation and region ranking for 3D reconstruction of reflective and uneven surface. In: IEEE International conference on image processing, IEEE, pp 4912–4916

A Robot Self-learning Grasping Control Method Based on Gaussian Process and Bayesian Algorithm

Yong Tao, Hui Liu, Xianling Deng, Youdong Chen, Hegen Xiong, Zengliang Fang, Xianwu Xie and Xi Xu

Abstract A robot self-learning grasping control method combining Gaussian process and Bayesian algorithm was proposed. The grasping gesture and parameters of the robot end-effector were adjusted according to the position and pose changes of target location to realize accurate grasping of the target. Firstly, a robot self-adaptive grasping method based on Gaussian process was proposed for grasping training in order to realize modeling and matching of position and pose information of target object and robot joint variables. The trained Gaussian process model is combined with Bayesian algorithm. The model was taken as priori knowledge and the semi-supervised self-learning was implemented in new grasping region so that posterior Gaussian process model was generated. This method omits the complex visual calibration process and inverse kinematics solves only with a small group of samples. Besides, when the environment of grasping changes, the previous learning experience can be used to perform self-learning, and adapt to the grasping task in the new environment, which reduces the workload of operators. The effectiveness of the robot self-learning grasping control method based on Gaussian process and Bayesian algorithm was verified through simulation and grasping experiment of UR3.

Keywords Gaussian process · Bayesian algorithm · Robot grasping
Semi-supervised self-learning

Y. Tao · H. Liu (✉) · Y. Chen · Z. Fang
Beihang University, Beijing 100191, China
e-mail: huiliu@buaa.edu.cn

X. Deng
Chongqing University of Science and Technology, Chongqing 401331, China

H. Xiong · X. Xie · X. Xu
Wuhan University of Science and Technology, Wuhan 430081, China

© Springer Nature Singapore Pte Ltd. 2018
S. Chen et al. (eds.), *Transactions on Intelligent Welding Manufacturing*,
Transactions on Intelligent Welding Manufacturing,
https://doi.org/10.1007/978-981-10-8330-3_5

1 Introduction

With maturation of robot technology in daily life, robots have entered people's life and production more and more frequently. In industrial production, it's a quite common application to operate manipulators to grasp target object such as transporting goods, systemizing articles and assembling parts which contain the most basic pick-and-place tasks. The grasp algorithm is also a major research hotspot in present robot orientation [1–6]. In operation of robot grasping articles, the position and pose of the grasped target object are usually not fixed. The self-adaptive grasping should be completed by adjusting grasping posture of robot end-effector according to position and pose of the target object.

According to different requirements for self-adaptive grasping of target object, some studies have added tactile sensors at end-effectors. Based on implemented encoding and decoding analysis according to contact information fed back by sensors, the grasping effects have been evaluated and motion parameters of the robot have been adjusted to complete favorable grasping [7–10]. In the Ref. [8], a probabilistic learning method for evaluating the stability of the grasp has been proposed. According to the tactile sensor feedback information, the performance of grasp is evaluated, so that objects can be regrasped before attempting to further manipulate them. In Refs. [9, 10], a hierarchical mechanism of two-step grasping is established. In the first step, based on tactile feedback, a grasp stability predictor is trained by supervised learning to predict performance of the grasp. In the second step, the parameters of grasp action are finely tuned according to the tactile sensor feedback. The methods above can improve adaptability of end-effector. However, it can only realize self-adaptive grasping nearby working region of the end-effector, and the posture adjustment by tactile perception is adverse to real-time grasping operation.

More studies have used visual sensors to obtain objects' position and pose to adjust the motion of robot and completed self-adaptive grasping of target object. The research methods of using visual sensor are divided into analytical methods and learning-based methods. Analytical methods use the image and depth information of the observed objects to model reshaping and point cloud segmentation, and then match a 3D model for each segmented object and analysis to obtain their pose information. Combined with the physical characteristics of each operation object and a grasp quality metric [11, 12], simulation would be performed in the physical simulator. Ultimately, suitable action would be chosen to perform the adaptive grasp of the objects [13–15]. However, the methods above have the following defects: firstly, complete model parameters are needed to pre-model the grasping objects which increase the workload. Secondly, it's difficult for the present depth and visual sensors to obtain complete model information and errors will easily occur during model matching process. The error exists between simulation result in simulator and actual operation result so that reliability of grasping algorithm will be greatly degraded. Thirdly, these methods need visual calibration for depth visual system of the robot. The traditional robot visual system calibration method has high requirements for professional knowledge and operating skills of working personnel [16–18], which consumes more time.

In the study of learning-based methods, the deep-learning technology is used. By constructing the deep-learning network, the mapping from images to action parameters of robot is established. The entire training process does not require any human intervention or model parameters and pre-modeling. Meanwhile, the calibration of camera is omitted, and it has stronger generalization ability and robustness. In Refs. [19, 20], manipulators collected data through self-supervised learning and trained a large Convolutional Neural Network (CNN) to generate the optimal grasping according to the visual information. In Ref. [21], CNN is trained with LSTM. The network uses unlabeled data to perform an end-to-end learning which established a predictor for estimating grasping effect. However, the application of deep-learning and neural network technology in learning-based methods requires a large number of training samples, and the training process is very time-consuming. Meanwhile, a long time to collect samples and training will increase wear and tear of hardware. How to get good learning effects through less training samples is a question worthy of study.

The visual servo system combines robot control and vision together without the need for pre-modeling and visual calibration [22, 23]. Indrazno [24] uses visual servo to achieve adaptive position control of a 7-DOF robot. Thomas [25] uses the visual servo to control the aircraft to complete the grasp action and the docking action. In Ref. [26], visual servo controller combined with reinforcement learning is applied to a mobile robot with a manipulator, which shows a good performance in robust grasping tasks. The use of visual servo can achieve the robot to adapt to the target object pose. However, the adjustment process of robot movement needs to analyze and deal with the image continuously, which increases the computational burden of the system and reduces the efficiency of the system.

In order to avoid solving the inverse solution and simplify the motion planning process, some studies teach robot to complete the grasping task by demonstration. In Ref. [27], the robot is taught to grasp different kind of objects by demonstration. The grasp type and the thumb position of each demonstration are recorded as the label of the corresponding grasp task. In Ref. [28], iterative learning is applied to the robot adaptive grasping control. In the learning process, adding manual adjustment of the robot's operating parameters to the demonstration, the robustness of the grasping control is increased. Ref. [29] proposed an object centred probabilistic volumetric model used to combine the multimodal data in the demonstration. The feature extracted by this method is proved to be useful for segmentation of the action phases and trajectory classification. In the process of demonstration, in order to generalize the executable tasks and improve the adaptability of the grasp task, it is particularly important to choose a mapping model from the observations to the motion parameters. As a Bayesian learning method, Gaussian process method (GPM) [30] can give robots the ability to learn the mapping function from the samples. Only small group of samples are needed to complete the training of the GPM and construct the nonlinear relation between the relevant variables. To solve the problem of physical Human-Robot interaction, Ghadirzadeh [31] uses GPM to establish the map between state-action pair and the variation of observations. In Ref. [32], in order to build a visual forward model, GPM is used to establish the mapping relation between motor commands and image observations. The researches above show that GPM is a powerful tool for non-parametric and non-linear regression.

In general, the requirements of environment for visual servo system are strict. This means that if the relative position of the robot and camera slightly changes, or if a grasping task is performed in a new and uncalibrated area, the accuracy of the grasp will be poor. Having this problem, in the application process whenever the environment changes, operators have to repeat the cumbersome calibration of visual system, which is time-consuming and increase the workload of operators. As for the GP method, under the circumstance of small sample size, it's not enough to rely only on GP probabilistic prediction model as it can obtain favorable generalization ability only nearby training samples. When the grasping environment changes or in a new region, the error also tends to increase. In order to adapt the GP model to the changing environment and to expand the scope of adaptive grasping, Bayesian method is adopted to combine with GPM as a self-learning algorithm.

A robot self-adaptive grasping algorithm based on Gaussian process was proposed. The position and pose information of target object obtained through visual grasping and corresponding robot joint variables were associated. It's only necessary to let the robot learn from demonstrated samples under small sample size which omitted calibration of robot visual system and inverse kinematics solving. Then the robot self-learning grasping control method based on Gaussian process and Bayesian algorithm was presented to do semi-supervised self-learning grasping in a new grasping region to generate the posteriori GP model, which improved adaptability of robot grasping. Grasping experiment of UR robot proved that robot self-learning grasping control method based on Gaussian process and Bayesian algorithm was of favorable effect.

2   Self-adaptive Grasping Based on Gaussian Process

2.1   Task Modeling

When executing a grasping task, the robot needs to match its own joint angle parameters and adjusts its motion according to position information of the target object. Self-adaptive grasping model of the robot to the target object is generalized as below:

$$f : o \rightarrow a \tag{1}$$

where o represents observations of target object, a is the corresponding joint coordinate, and f is mapping function from observation variable of the target object to joint coordinate.

It is assumed that is $X = \{x_1, x_2, \ldots, x_n\}$ the sample set obtained through demonstration, where $x_i = [a_i, o_i]^T$ represents sample vector consisting of joint variables and observations. Robot self-adaptive grasping process lies in learning from sample set X to obtain mapping function f so that the robot can obtain the corresponding joint coordinate a according to the observation o.

2.2 System Description

As shown in Fig. 1, the right part is experiment platform, left part is UR manipulator of six degrees of freedom, and an industrial camera is fixed right above the experimental platform. The camera optic axis is vertical with experimental platform, and the object is located within visible scope of the camera on the experimental platform. Two coordinate systems in the figure are respectively $\{B\}$ manipulator basic coordinate system and $\{T\}$ tool coordinate system.

Within a certain training region, the target object is arbitrarily placed, the industrial camera acquires pixel coordinates of the target location. The demonstration is then given, the most appropriate robot joint angle is selected so that the robot end-effector can execute grasping task accurately. The robot joint and observation of pixel coordinates of the target location are established as the sample sets. With information from training sets, Gaussian process model is used to establish mapping relation between the two sets. In the experiments, the robot completes the grasping task through robot joint angle correspondingly predicted through Gaussian process according to position information of target object obtained by the camera.

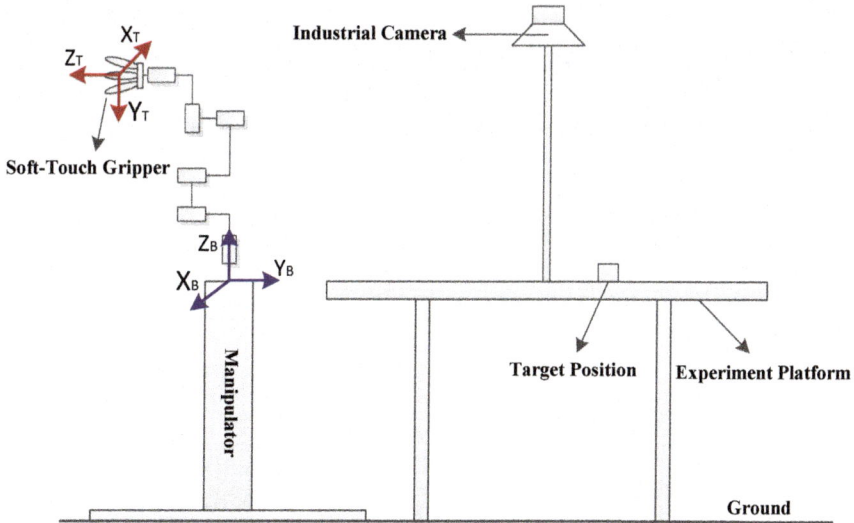

Fig. 1 The platform of manipulator grasping

2.3 *Gaussian Process Model*

GPM is a nuclear learning machine with probabilistic meaning, which can give a probabilistic interpretation of the predicted output. GPM is based on the assumption that the observations and predictions are subject to a joint normal distribution, then the posterior distribution of the predictions would be obtained by solving the covariance matrix of the observations and the input of the training set. GPM has been applied to the regression and classification problems successfully [10, 11]. The robot self-adaptive grasping method based on Gaussian process is as shown in Fig. 2. The method has omitted calibration of visual system and inverse kinematics solving. The robot needs to learn from samples to obtain parameters of Gaussian process model.

Before data are obtained, it's assumed that joint variable and observational variable of the target object comply with Gaussian distribution with mean value μ and covariance matrix K:

$$h \sim (\mu, K) \tag{2}$$

In the equation, $h = [a, o]^T$ represents vector consisting of observation and joint variables. Sample set $X = \{x_1, x_2, \ldots, x_n\}$ obtained through demonstration includes measurement noise, and then:

$$x_i = h + \varepsilon \tag{3}$$

In the equation, ε represents Gaussian noise with mean value 0 and variance σ_n^2.

Posteriori distribution of multidimensional variables obtained through sample set X is also Gaussian distribution:

$$p(h|X, \theta) = N(\mu, K + \sigma_n^2 I) \tag{4}$$

where $\theta = \{\mu, K, \sigma_n^2\}$. For sample set X, its marginal likelihood function is:

Fig. 2 Adaptive grasping based on Gaussian process

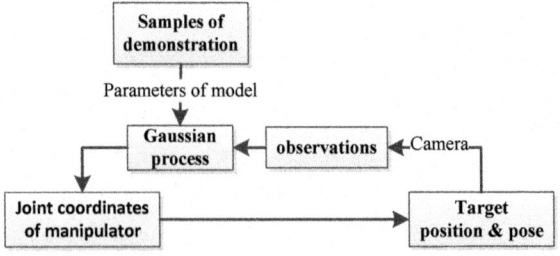

$$p = (X|H, \theta) = \prod_{i=1}^{n} p(x_i|h_i, \theta) = \prod_{i=1}^{n} p(x_i|h_i)p(h_i|\theta) = P \tag{5}$$

where $P = \prod_{i=1}^{n} \frac{1}{\sqrt{2\pi|K+\sigma_n^2 I|}} \exp\left(-\frac{\tau_i}{2}\right)$, and $\tau_i = (x_i - \mu)^T (K + \sigma_n^2 I)^{-1} (x_i - \mu)$

Partial derivatives of mean value vector and covariance matrix of the model are respectively solved through Eq. (5), derivatives are set as 0. The maximum likelihood estimation values of mean value vector and covariance matrix of Gaussian process can be obtained respectively as below:

$$\mu = \frac{\sum_{i=1}^{n} x_i}{n} \tag{6}$$

$$(K + \sigma_n^2 I) = Cov([x_1, x_2, \ldots, x_n]^T) \tag{7}$$

2.4 Prediction of Joint Variables

The following is obtained by blocking vector and matrix of Gaussian process:

$$\begin{bmatrix} a \\ o \end{bmatrix} \sim N\left(\begin{bmatrix} \mu_a \\ \mu_o \end{bmatrix}, \begin{bmatrix} K_{aa} + \sigma_n^2 I & K_{ao} \\ K_{oa} & K_{oo} + \sigma_n^2 I \end{bmatrix}\right) \tag{8}$$

The robot obtains observation information o^* of the target object from the camera, and then corresponding conditional probability distribution of joint angle a^* is:

$$p(a^*|o^*) = N\left(\mu_a^*, K_{aa}^*\right) \tag{9}$$

where $\mu_a^* = \mu_a + K_{ao}\left(K_{oo} + \sigma_n^2 I\right)^{-1}(o^* - \mu_o)$, $K_{aa}^* = K_{aa} - K_{ao}\left(K_{oo} + \sigma_n^2 I\right)^{-1} K_{oa}$.

μ_a^* is mean value of joint angle matching new target location and it's corresponding to maximum probability of Gaussian distribution. K_{aa}^* is covariance matrix of Gaussian distribution and it represents uncertainty of prediction result. Grasping can be completed by the robot at maximum probability by driving robot joint to reach μ_a^*.

As it's not necessary to do visual system calibration and inverse kinematics solving, Gaussian process directly associates robot joint variables and observational variables of the target object. According to new observations, the robot joint angle corresponding to position of the target object is predicted.

3 Semi-supervised Self-learning Grasping Based on Bayesian Algorithm

GP method performs well in the training region with only small dataset. The method is of decent generalization ability. However, some errors may exist in this method when it used for grasping beyond training region. In order to expand effective grasping scope of manipulator, the Bayesian algorithm is adopted. The GP model trained before is used as priori model which is then added into semi-supervised learning process. New training samples are collected after grasping training in new adjacent region through robot self-learning grasping so as to update probability distribution of the whole Gaussian process. The posteriori probability model is obtained.

As shown in Fig. 3, target position is randomly selected in new training region as input of GP model. Based on Gaussian self-adaptive strategy discussed in Sect. 2, the GP model is used as priori model to generate robot joint angle parameters. In supervised learning process, the solution of the forward kinematic of the manipulator is solved according to the joint angle. Then a posture evaluation mechanism is used for feedback and fine adjustment of joint angle. After a relatively reasonable terminal grasping posture is obtained, the manipulator will try grasping, and the actual grasping position of end-effector is acquired. On the one hand, joint angle parameters and corresponding actual grasping position are taken as new samples in training region to update the training set. On the other hand, according to actual grasping position and target location, grasping evaluation mechanism is used for evaluation.

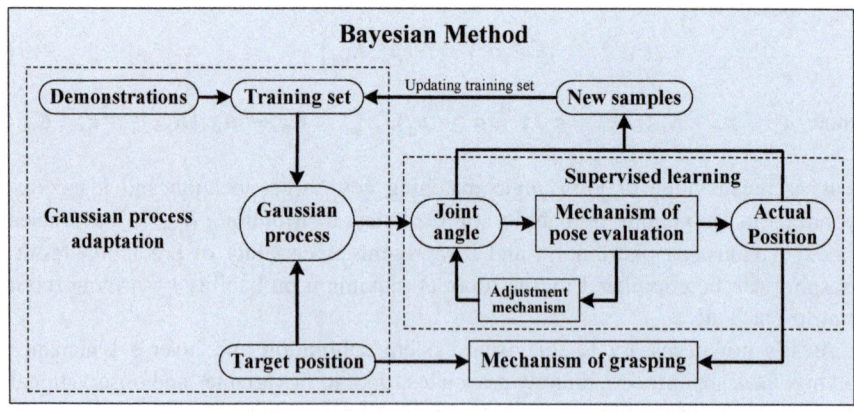

Fig. 3 Semi-supervised self-learning control method based on Bayesian method

3.1 Semi-supervised Learning

During demonstrated sampling process, collected sample posture is set as fixed posture. The gripper executes grasping from up to down in the direction vertical to plane of the experimental platform as shown in Fig. 4. Z_T axis of tool coordinate system is parallel to and in reverse direction to Z_B axis. Correspondingly, $X_T O_T Y_T$ plane is parallel to $X_B O_B Y_B$ plane. H is the height of grasping location of the target object under $\{B\}$ robot reference coordinate system. During semi-supervised learning process, robot joint angle is generalized according to target location through Gaussian process. In order to guarantee effectiveness of successful grasping, final grasping posture of end-effector is made to approach posture in demonstration samples as much as possible. Meanwhile, the height of grasping location should approach height of end-effector in demonstration samples as much as possible. Adjustment is completed through two iterative loops, and concrete operation is as below:

According to pixel coordinate of objects in the new adjacent training region, the self-adaptive grasping method based on Gaussian process in Sect. 2 is used to predict joint angle vector μ_a^* with maximum successful grasping rate. The D-H modeling of the manipulator is implemented and joint angle vector μ_a^* is substituted into forward solution formula of robot kinematics to obtain coordinate transformation matrix.

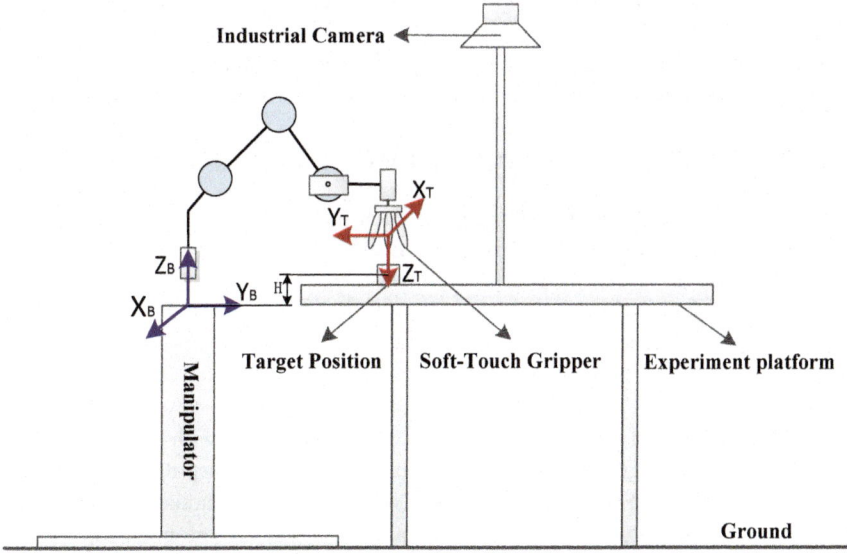

Fig. 4 The pose of end-effector

$$T = \begin{bmatrix} r_{11} & r_{12} & r_{13} & p_x \\ r_{21} & r_{22} & r_{23} & p_y \\ r_{31} & r_{32} & r_{33} & p_z \\ 0 & 0 & 0 & 1 \end{bmatrix}$$

θ_z is used to express relationships between Z_T axis in tool coordinate system and axes of reference coordinate system. P represents displacement of origin of tool coordinate system in direction of Z_B axis under reference coordinate system.

$$\theta_z = \begin{bmatrix} \frac{\pi}{2} - \arccos(r_{13}) \\ \frac{\pi}{2} - \arccos(r_{32}) \\ \pi - \arccos(r_{33}) \end{bmatrix}; \quad P = p_z - H$$

where $\theta_{zi} \in \left[-\frac{\pi}{2}, \frac{\pi}{2}\right]$, $i = 0, 1$; $\theta_{zi} \in [0, \pi]$, $i = 2$.

According to obtained θ_z, loop iteration A is implemented and robot joint angle is adjusted.

$$\mu_a = \mu_a + w\theta_z^T$$

where μ_a is the robot joint angle, and w is correction coefficient matrix.

Termination conditions of iterative loop A are:

$$\begin{cases} \text{Execute loop A and enter loop B} & |\theta_{zi}| < \delta, i = 0, 1, 2 \\ \text{Execute loop A} & \text{else} \end{cases}$$

After termination of iterative loop A, joint angle parameters with reasonable grasping posture are obtained to do iterative loop B so as to adjust robot joint angle.

$$\mu_a = \mu_a + w'P$$

where w' is correction coefficient vector of each joint angle.

Termination conditions of iterative loop B are:

$$\begin{cases} \text{Goto loop A} & |P| < h \text{ and } |\theta_{zi}| \geq \delta, i = 0, 1, 2 \\ \text{Execute loop B,} & |P| \geq h \\ \text{End loop B,} & \text{else} \end{cases}$$

After termination of iterative loop B, joint angle parameters with both reasonable grasping posture and grasping height are obtained. Execute grasping operation and observe the grasping position of the gripper which is taken as new training sample together with joint angle parameters. The training sets of GP are updated.

3.2 Bayesian Algorithm

The Bayesian algorithm has been proposed to do modeling of grasping in new training region. Main idea of Bayesian algorithm can be expressed by the following equation:

$$p(\theta|X) = \frac{p(X|\theta) \cdot p(\theta)}{p(X)}$$

where $p(\theta)$ is probability distribution of priori model, X is training sample, $p(\theta|X)$ is probability distribution of posteriori model and $p(X)$ is boundary likelihood which is solved through the following equation:

$$p(X) = \int p(X|\theta)P(\theta)\,d\theta$$

Firstly, target position (x_d, y_d) is randomly selected in new training region as input of GP model, and its output is predicted joint angle $\boldsymbol{\mu}_a^*$. Through posture evaluation and mechanism of adjustment, actual position (x_a, y_a) of gripper is distributed nearby the target position, but it will not necessarily coincide with target position. Actual position (x_a, y_a) and corresponding joint angle parameter are added into training set as new samples. After enough samples in new training region are obtained, posteriori Gaussian distribution containing new training region will be established. The grasping evaluation mechanism is established according to actual grasping location and joint angle parameters generated through posteriori model. The termination conditions of posteriori model are defined. The evaluation function is designed as below:

$$r = cV_r e^{-\lambda\Delta d} + V_r', \Delta d = \sqrt{(x_a - x_d)^2 + (y_a - y_d)^2}$$

where c represents whether manipulator goes through posture adjustment, $c = 0$ means yes and $c = 1$ means no, and extra bonus will be obtained when the robot doesn't need posture adjustment. V_r is the reward value when deviation Δd is 0, λ is a sufficiently large parameter which can guarantee that $e^{-\lambda\Delta d}$ converges to 0 when Δd is great enough. V_r' is the reward value for a newly added sample.

Termination condition of model training is as below:

When $R = \sum_{i=1}^{n} r_i > \kappa$ is satisfied, model training is terminated.

The self-learning algorithm is shown in following as Algorithm 1.

Algorithm 1

Input: sample set of demonstration: $X = \{x_1, x_2, \ldots, x_n\}$; observation space O;

cumulative reward $R = 0$; distribution of prediction: $h = [a, o]^T \sim (\mu, K)$

for $t = 1, 2, \ldots$ do

$\quad \forall o \in O: a = \mu_a + K_{ao}(K_{oo} + \sigma_n^2 I)^{-1}(o^* - \mu_o)$;

$\quad$ Loop A: $(\theta_z, P) = T(a)$

$\qquad$ if $\forall i \in \{1,2,3\}: |\theta_{zi}| < \delta$, then

$\qquad\qquad$ end Loop A

$\qquad$ else

$\qquad\qquad a = a + w\theta_z^T$

$\quad$ Loop B: $(\theta_z, P) = T(a)$

$\qquad$ if $|P| < h$ and $\exists i \in \{1,2,3\}: |\theta_{zi}| > \delta$, then

$\qquad\qquad$ Goto Loop A

$\qquad$ else if $|P| \geq h$, then

$\qquad\qquad a = a + w'P$

$\qquad$ else

$\qquad\qquad$ end Loop B

$\quad$ execute a, obtain o' in O, then add $x_{n+t} = (o', a)^T$ into X;

$\quad$ update the posterior distribution of h: $p(h|X, \theta) = N(\mu, K + \sigma_n^2 I)$, where

$\quad \mu = \frac{\sum_{i=1}^{n+t} x_i}{n+t}$, $(K + \sigma_n^2 I) = Cov\left(\left[x_1, x_2, \ldots, x_{n+t}\right]^T\right)$;

$\quad r(o, o') =$ the reward of executing a in O, $R = R + r$;

$\quad$ if $R = \sum_{i=1}^{t} r_i > \kappa$

end for

Output: new sample set X, new distribution of prediction h

4 Simulation and Experiment

Simulation and experimental object is visual grasping platform based on UR3. UR3 contains 6 joint axes with high flexibility and operability. After new observational variable o^* is obtained, the trained GP model is used to calculate joint a^* which the robot needs to reach so as to realize robot self-adaptive grasping of the target object.

4.1 Simulation of Control Method

Gaussian self-adaptive grasping

Simulation uses Robotic Toolbox in MATLAB. Firstly, demonstration of the robot is performed, corresponding joint angle is input so that the gripper will grasp the object in vertical direction. Meanwhile, the certain grasping height is ensured so that gripper can grasp the target successfully. The robot joint angle and pixel coordinate of the gripper after coordinate transformation are recorded as training samples. They are input into the training sample set of GP model. In the simulation, 13 pixels' coordinates are uniformly collected in training region, and training data are shown in Table 1. After training samples are obtained, maximum likelihood estimations of mean value vector μ and covariance matrix $(K + \sigma_n^2 I)$ of GP are obtained through Eqs. (6) and (7). New observation and robot joint comply with the same probability distribution. When there is new observation input o^*, joint angle μ_a^* which allows the robot complete grasping at maximum probability can be solved through Eq. (9).

Distribution of training sample points and test sample points inside pixel plane is as shown in Fig. 5. The left green dotted box expresses the training region and the right red dotted box expresses untrained one. Blue points are training samples uniformly collected, and the other points are test samples. The results of grasping are judged according two rules. First, the distance from gripper and target location should be small enough. When the distance is smaller than 10 mm, it can be considered as a successful grasp. Second, height of gripper should be appropriate. In the simulation, when height of gripper is equal to 69 ± 4 mm, it can be considered that grasping is successful. Based on the two rules, the green points in the figure are successful samples. The yellow points express samples with too large deviation of grasping distance. The red points express samples with inappropriate grasping height.

As shown in Fig. 5, most test points in the training region can be successfully grasped. It can be seen that under small sample size, Gaussian process model has considerable performance in the training region. However, test points nearby boundary of training region and those beyond training region almost failed to grasp, and grasping effect is unsatisfying. Therefore, Gaussian process model relies on priori samples and has poor performance under the circumstance in which there are no priori samples.

Table 1 Training data from demonstration

Num	Observations		Corresponding joint angle (rad)					
	Pixel x	Pixel y	Base	Shoulder	Elbow	Wrist 1	Wrist 2	Wrist 3
1	239	88	−0.134	−1.730	−2.022	−0.965	1.573	0.007
2	347	88	0.003	−1.661	−2.103	−0.951	1.574	0.145
3	455	87	0.149	−1.609	−2.161	−0.944	1.575	0.291
4	455	191	0.130	−1.768	−1.976	−0.970	1.574	0.271
5	347	192	0.003	−1.810	−1.920	−0.985	1.573	0.145
6	239	192	−0.118	−1.868	−1.841	−1.008	1.572	0.023
7	239	297	−0.105	−2.004	−1.639	−1.073	1.571	0.036
8	347	297	0.003	−1.953	−1.717	−1.045	1.572	0.144
9	454	296	0.116	−1.917	−1.772	−1.026	1.573	0.257
10	401	244	0.062	−1.861	−1.852	−1.003	1.573	0.204
11	293	245	−0.055	−1.907	−1.785	−1.024	1.572	0.086
12	293	140	−0.062	−1.767	−1.977	−0.973	1.573	0.079
13	401	140	0.071	−1.712	−2.045	−0.958	1.574	0.212

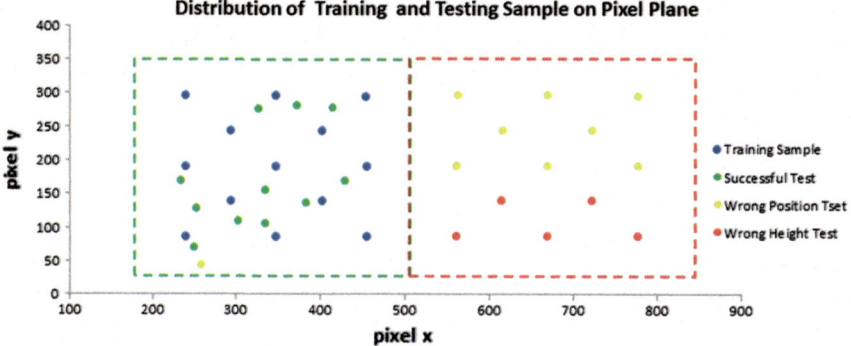

Fig. 5 Distribution of training and testing samples on pixel plane

Self-learning grasping

It's assumed that in new training region, probability distribution which observations and robot joint angles subject to is identical with that in the previous one. The trained GP model is taken as priori model, 12 test sample points are uniformly collected in new training region. The obtained joint angle and grasping errors are shown in Table 2.

Table 2 shows that in new training region, all test points failed to grasp. The average grasping error is 25.4 mm and maximum deviation reaches as high as 49 mm. Grasping height of more than 1/3 test points do not meet requirement. Then, self-learning is implemented in the new training region, and steps are as below:

Table 2 Test samples in new training region

No.	Pixel x	Pixel y	Height	Error	Base	Shoulder	Elbow	Wrist 1	Wrist 2	Wrist 3
1	560	88	63.3	15.26	0.253	−1.568	−2.232	−0.912	1.576	0.395
2	668	88	60.2	29.77	0.378	−1.518	−2.299	−0.895	1.577	0.521
3	777	87	56.5	49.08	0.505	−1.465	−2.369	−0.877	1.578	0.647
4	776	191	65.7	39.40	0.503	−1.610	−2.176	−0.924	1.577	0.645
5	668	192	67.8	21.63	0.378	−1.662	−2.107	−0.942	1.576	0.520
6	560	192	69.2	8.86	0.252	−1.713	−2.041	−0.960	1.575	0.394
7	561	297	70.5	11.38	0.253	−1.859	−1.847	−1.007	1.574	0.394
8	668	297	70.7	22.44	0.377	−1.809	−1.914	−0.990	1.575	0.519
9	776	296	70.4	38.83	0.503	−1.757	−1.982	−0.972	1.576	0.644
10	722	244	69.3	28.64	0.440	−1.709	−2.045	−0.957	1.576	0.582
11	615	245	70.2	14.02	0.316	−1.761	−1.977	−0.975	1.575	0.458
12	614	140	65.8	17.57	0.315	−1.615	−2.170	−0.927	1.576	0.457

1. Test samples are selected in new training region, and robot joint angles are obtained according to previous GP model.
2. In simulation, forward kinematic solution is solved according to joint angles. Iterative loop will be performed and joint angles will be adjusted if it's judged the posture or height is inappropriate. Then, go to the next step.
3. Corresponding pixel coordinates of the gripper are observed and recorded. which will be taken as new samples together with newly obtained robot joint angles. Input new samples into the training set of GP so as to update the whole distribution.
4. Reward value is calculated and updated, and it's judged whether self-learning training is completed. If it's judged that it's not completed, then return to 1, or otherwise, end the loop.

During learning process, the generated new samples are input into training set of GP so as to update the whole distribution. As shown in Fig. 6, the first row is distribution of predicted values of 6 joint angles in new training region when new training samples are not added. The second row is distribution after 6 new samples are added. The third row is distribution after 12 new samples are added. With increase in the number of new samples, distribution of robot joint angles changes. Reward value reaches threshold value through test sample point collection and updating of posteriori model for 12 times so that self-learning training is completed.

After self-learning of grasping, 100 sample points ate randomly tested in the new region, and distribution of test results is shown in Fig. 7. In the new training region, 83% test sample points can be successfully grasped. 14% test sample points failed to grasp because of deviation of position, where distance errors of 11% samples are within 10~15 mm. 3% samples failed to grasp because their grasping height is wrong. On the whole, posteriori GP model after self-learning has favorable effect in

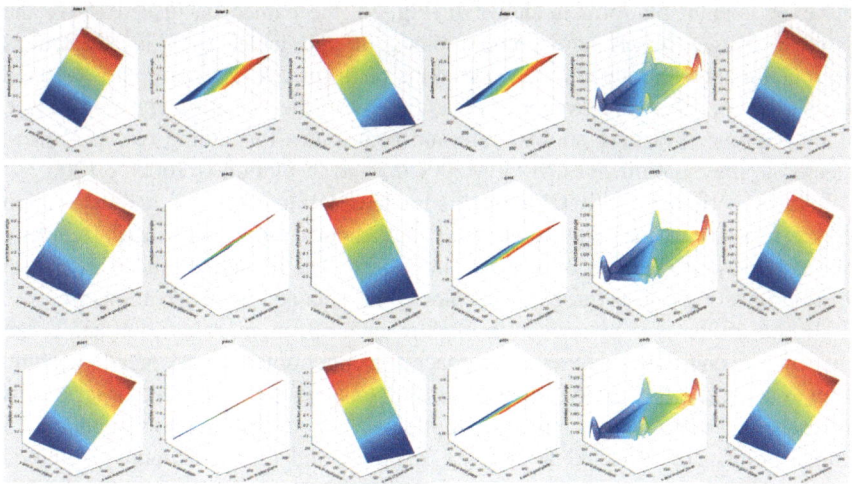

Fig. 6 Variation of the distribution of manipulator's joints

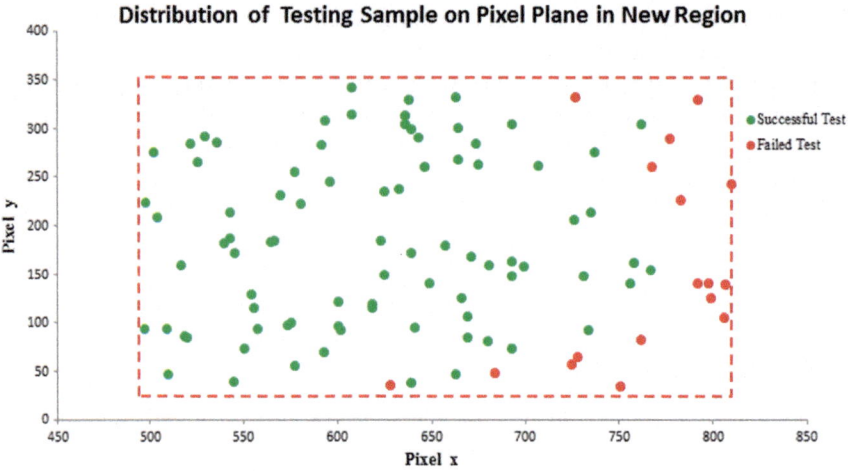

Fig. 7 Distribution of test samples in new region after self-learning

the new training model. Furthermore, most fail samples are distributed far away from training sample points. This conforms to characteristics of GP model. In the region with concentrated samples, the prediction performance is good, and in the region with sparse samples or being far away from samples, the performance of prediction tends to be poor.

4.2 Experiment

The experimental platform is shown in Fig. 8a, the camera is placed above the platform. Target object is a colored block on the table. UR3 is selected as manipulator and is arranged at left side of the platform. SRT soft gripper is selected as end-effector. The soft gripper is of adaptivity to the shape of grasping object, which improves grasping success rate and contributes to training effect. Camera selected in the experiment is MV-EM200C/M and resolution is 1600×1200.

In the experiment, firstly GP model is trained. Profile extraction and calculation of central point of the target object are implemented through industrial camera, image processing interface is as shown in Fig. 8b. The pixel coordinates of central position of target object (green wood block) are obtained as observations $o(x, y)$.

Demonstration is implemented through manual operation, robot joint angles are adjusted, the end-effector is set to approach the target object in reasonable posture, robot joint angle a is recorded when successful grasping is ensured, and observations and joint angles are taken as training samples of GP model. Similar to simulation, 13 sample points are collected as training samples, and then maximum likelihood

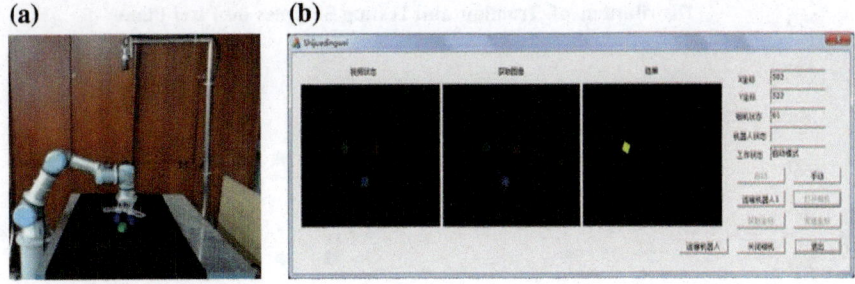

Fig. 8 Experiment of grasping task: **a** experimental platform; **b** acquiring observation variables of target object

estimation values of mean value vector $\boldsymbol{\mu}$ and covariance matrix $\left(\boldsymbol{K} + \sigma_n^2 \boldsymbol{I}\right)$ of GP are obtained through Eqs. (6) and (7). Training data are shown in Table 3.

Green wood blocks are randomly placed in training region, and 12 samples points are tested. The distribution of training samples and test samples are shown in Fig. 9. Blue points are training samples which are uniformly collected in training region. Green points are successful samples and red points are failed samples. Similar to the simulation, most test samples can be successfully grasped, several failed samples are located at edges of training region. In general, in the training region, self-adaptive grasping of GP has decent performance result. However, similar to simulation, beyond the training region, previous GP model is tried for estimation to generate joint angles result, and nearly all test samples failed to grasp. GP model has poor performance in the region without prior knowledge.

Table 3 Training data from demonstration

No.	Observations		Corresponding joint angle (rad)					
	Pixel x	Pixel y	Base	Shoulder	Elbow	Wrist 1	Wrist 2	Wrist 3
1	240	455	−0.092	−2.220	−1.239	−1.271	1.573	6.188
2	346	455	0.001	−2.171	−1.324	−1.233	1.575	6.281
3	453	454	0.096	−2.136	−1.382	−1.208	1.577	0.093
4	453	559	0.088	−2.304	−1.092	−1.331	1.576	0.085
5	347	559	0.002	−2.340	−1.026	−1.362	1.574	6.281
6	240	559	−0.083	−2.395	−0.925	−1.410	1.572	6.196
7	241	667	−0.076	−2.650	−0.455	−1.620	1.571	6.204
8	347	670	0.002	−2.564	−0.620	−1.530	1.573	6.282
9	454	663	0.082	−2.514	−0.701	−1.513	1.575	0.078
10	400	611	0.043	−2.418	−0.883	−1.427	1.574	0.040
11	294	612	−0.039	−2.469	−0.787	−1.474	1.573	6.240
12	293	507	−0.043	−2.276	−1.142	−1.312	1.574	6.236
13	400	507	0.047	−2.233	−1.218	−1.277	1.576	0.043

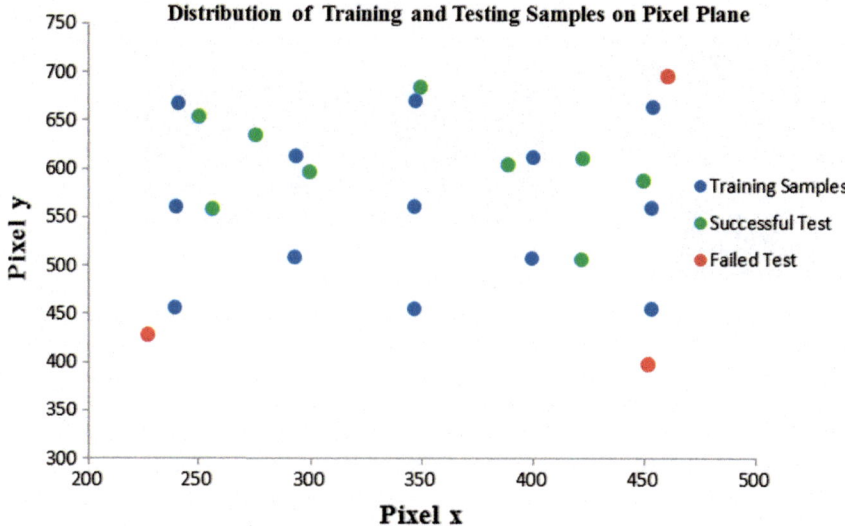

Fig. 9 Distribution of training and testing samples on pixel plane

Self-learning training is carried out according to simulation steps, subsequently, green blocks are randomly placed in new training region, and 250 samples are tested, and test results are shown in Table 4. Among 250 experimental tests, the number of successful grasping is 194 occupying 77.6% which is slightly lower than in simulation. That is possibly caused by errors of camera and precision of manipulator. Among unsuccessful samples, samples with deviation of grasping location occupy 12.4% which constitute the main type of unsuccessful grasping. Blocks are successfully grasped but drop during movement process of manipulator, it belongs to unsafe grasping situation, and this occupies 6.4%. The situation of nothing grasping or gripper touching experimental platform due to wrong grasping height occupies 3.6%. In general, after self-learning, posteriori GP has favorable performance result in the new training region.

Table 4 Grasping test in experiment

The result of grasping	No.	Proportion
Successful grasping	194	77.6%
Grasp wrong position	31	12.4%
Unstable	16	6.4%
Grasp wrong height	9	3.6%

5 Conclusion

Robot self-adaptive grasping strategy based on Gaussian process was proposed. The pose and position information of the target object obtained through visual grasping and corresponding robot joint variables were associated. On the condition that only a small sample size was needed, the robot was made to learn from artificial demonstration samples. The robot self-learning grasping method based on Gaussian process and Bayesian algorithm was presented. The robot could use previous Gaussian process model as priori model to implement self-learning grasping in the region without prior knowledge, training scope of Gaussian process model was expanded and adaptability of robot grasping was improved. This method omits the complex and time-consuming calibration of visual system. Besides, choosing similar results with the demonstration samples, it does not need to solve the inverse kinematic and the optimal solution which improves the efficiency of grasping. When grasping environment changes, the previous learning experience can be used as priori knowledge. The self-learning is able to complete the grasping task in new environment in relative high success rate without repeating calibration process, which reduces the workload of operators.

References

1. Kroemer OB, Detry R, Piater J et al (2010) Combining active learning and reactive control for robot grasping. Robot Auton Syst 58(9):1105–1116
2. Lenz I, Lee H, Saxena A (2013) Deep learning for detecting robotic grasps. Int J Robot Res 34(4–5):705–724
3. Manti M, Hassan T, Passetti G et al (2015) A bioinspired soft robotic gripper for adaptable and effective grasping. Soft Robot 2(3):107–116
4. Levine S, Finn C, Darrell T et al (2016) End-to-end training of deep visuomotor policies. J Mach Learn Res 17(39):1–40
5. Zeng A et al (2017) Multi-view self-supervised deep learning for 6D pose estimation in the amazon picking challenge. In: Proceedings of the IEEE international conference on robotics and automation, IEEE, Singapore, pp 1386–1383
6. Garcia-Sillas D et al (2016) Learning from demonstration with Gaussian processes. In: IEEE conference on mechatronics, adaptive and intelligent systems, IEEE, Hermosillo, pp 1–6
7. Shimojo M, Namiki A, Ishikawa M et al (2004) A tactile sensor sheet using pressure conductive rubber with electrical-wires stitched method. IEEE Sens J 4(5):589–596
8. Bekiroglu Y, Laaksonen J, Jorgensen JA et al (2011) Assessing grasp stability based on learning and haptic data. IEEE Trans Rob 27(3):616–629
9. Dang H, Allen PK (2014) Stable grasping under pose uncertainty using tactile feedback. Auton Robots 36(4):309–330
10. Chebotar Y et al (2016) Self-supervised regrasping using spatio-temporal tactile features and reinforcement learning. In: IEEE/RSJ international conference on intelligent robots and systems, IEEE, Daejeon, pp 1960–1966
11. Haschke R et al (2005) Task-oriented quality measures for dextrous grasping. IEEE international symposium on computational intelligence in robotics and automation. IEEE, Espoo, pp 689–694

12. Roa MA, Suárez R (2015) Grasp quality measures: review and performance. Auton Robots 38 (1):65–88
13. Miller AT et al (2003) Automatic grasp planning using shape primitives. In: Proceedings of the IEEE international conference on robotics and automation, vol 2, IEEE, pp 1824–1829
14. Ciocarlie MT, Allen PK (2009) Hand posture subspaces for dexterous robotic grasping. Int J Robot Res 28(7):851–867
15. Li Y, Saut J, Pettré J et al (2015) Fast grasp planning using cord geometry. IEEE Trans Rob 31(6):1393–1403
16. Zhang Z (2000) A flexible new technique for camera calibration. IEEE Trans Pattern Anal Mach Intell 22(11):1330–1334
17. Wang Y, Liu CJ, Ren YJ et al (2009) Global calibration of visual inspection system based on universal robots. Opt Precis Eng 17(12):3028–3033
18. Tsai RY, Lenz RK (1989) A new technique for fully autonomous and efficient 3D robotics hand/eye calibration. IEEE Trans Robot Autom 5(3):345–358
19. Levine S et al (2016) Learning hand-eye coordination for robotic grasping with largescale data collection. International symposium on experimental robotics. Springer, Berlin, pp 173–184
20. Pinto L et al (2016) Supersizing self-supervision: learning to grasp from 50 k tries and 700 robot hours. In: Proceedings of the IEEE international conference on robotics and automation, IEEE, Stockholm, pp 3406–3413
21. Finn C et al (2017) Deep visual foresight for planning robot motion. In: Proceedings of the IEEE international conference on robotics and automation, IEEE, Singapore, pp 2786–2793
22. Hutchinson S, Hager GD, Corke PI (1996) A tutorial on visual servo control. IEEE Trans Robot Autom 12(5):651–670
23. Chaumette F, Hutchinson S (2006) Visual servo control. I. basic approaches. IEEE Robot Autom Mag 13(4):82–90
24. Siradjuddin I et al (2012) A position based visual tracking system for a 7 DOF robot manipulator using a kinect camera. The 2012 international joint conference on neural networks. IEEE, Brisbane, pp 1–7
25. Thomas J et al (2014) Toward image based visual servoing for aerial grasping and perching. In: Proceedings of the IEEE international conference on robotics and automation, IEEE, Hong Kong, pp 2113–2118
26. Wang Y, Lang H, Silva CW (2010) A hybrid visual servo controller for robust grasping by wheeled mobile robots. IEEE/ASME Trans Mechatron 15(5):757–769
27. Lin Y et al (2014) Grasp planning based on strategy extracted from demonstration. IEEE/RSJ international conference on intelligent robots and systems. IEEE, Chicago, pp 4458–4463
28. Sauser EL, Argall BD, Metta G et al (2012) Iterative learning of grasp adaptation through human corrections. Robot Auton Syst 60(1):55–71
29. Faria DR, Martins R, Lobo J et al (2012) Extracting data from human manipulation of objects towards improving autonomous robotic grasping. Robot Auton Syst 60(3):396–410
30. Rasmussen CE, Williams CK (2004) Gaussian processes in machine learning. Lect Notes Comput Sci 3176:63–71
31. Ghadirzadeh A et al (2016) A sensorimotor reinforcement learning framework for physical human-robot interaction. IEEE/RSJ international conference on intelligent robots and systems. IEEE, Daejeon, pp 2682–2688
32. Ghadirzadeh A et al (2014) Learning visual forward models to compensate for self-induced image motion. The 23rd IEEE international symposium on robot and human interactive communication. IEEE, Edinburgh, pp 1110–1115

Real-Time Implementation of a Joint Tracking System in Robotic Laser Welding Based on Optical Camera

Qi Zhang, Shanglei Yang, Haobo Liu, Chaojie Xie, Yaming Cao and Yuan Wang

Abstract Robotic laser welding has been widely applied to industries due to its high flexibility and productivity. However, there are still some constrain such as heat induced deformation and inevitable fixture errors, which can affect the laser beam positioning to deviate from joint center, and that will lead to poor welding quality. In order to ensure high quality welding, a joint tracking system is needed to track the joint center in real time to keep the focus of laser beam following the weld joint consistently. This paper introduces laser welding, describes composition of a real-time joint tracking system which mainly includes image acquisition part, image process and analysis part, and motion control part, reviews relevant investigations of joint tracking system and algorithm. Although there are successful applications in real-time joint tracking, the systems and algorithms can only be used in limited situations. So, future research can be focused on problems such as the promotion of system performance, the commercial solutions for joint tracking, etc.

Keywords Laser welding · Robotic welding · Real-time · Joint tracking

1 Introduction

1.1 Problem Description

Compared with manual welding which is time consuming and inefficient, robotic welding has high flexibility and automation. At the same time, laser welding can realize deep penetration and improve mechanical properties due to a narrow heat

Q. Zhang · S. Yang (✉) · H. Liu · C. Xie · Y. Cao · Y. Wang
School of Materials Engineering, Shanghai University of Engineering Science,
Shanghai 201620, China
e-mail: yslei@126.com

S. Yang
Shanghai Collaborative Innovation Center for Laser Advanced Manufacturing Technology,
Shanghai University of Engineering Science, Shanghai 201620, China

© Springer Nature Singapore Pte Ltd. 2018
S. Chen et al. (eds.), *Transactions on Intelligent Welding Manufacturing*,
Transactions on Intelligent Welding Manufacturing,
https://doi.org/10.1007/978-981-10-8330-3_6

affected zone compared with other welding techniques e.g. arc welding [1, 2]. So, there would be more application space for robotic laser welding considering advantages of both robotic welding and laser welding.

However, there is only limited success in many welding applications because of many robotic laser welding systems can't adapt to real-time changes of joint or environment such as thermal distortion from the intense heart of the laser beam, fixture errors or improper preparation of the weld joint [3, 4]. These variations will lead to laser beam wandering off the joint and result in bad quality e.g. lack of penetration, unacceptable welds, and reducing of heat efficiency largely [5, 6], such as Fig. 1. Thus, in order to compensating these variations in changeable environment in real time and getting good welding quality, joint tracking is needed to keep laser beam following the weld joint in real-time consistently.

1.2 Related Work

There have been many investigations about real-time joint tracking. Reference [3] proposed a method for seam-tracking which was based on infrared sensors, the molten pool infrared images are got through camera at first, then, thermal distribution of joint is analyzed, finally, a dynamic visual model which can measure the offset between laser-beam focus and weld-seam center was established. A laser welding experimental platform for burred joint welding is introduced in [7], they proposed a novel vision sensor system, images of burred joint can be acquired accurately based on their system and the final quality of welding is good. Reference [8] proposed an automatic seam tracking system which used for multi-pass arc welding, and an algorithm which used for image processing was also proposed. The results showed that the system and algorithm have good performance for image processing and seam tracking. Reference [9] proposed an algorithm for real time

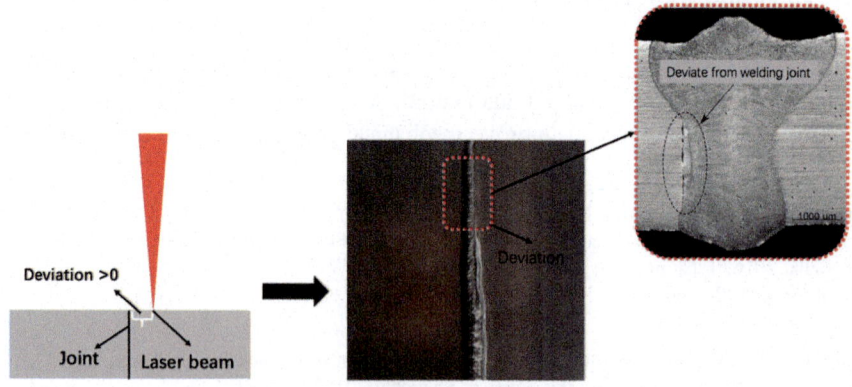

Fig. 1 Deviation between laser beam port and joint center leads to poor quality

seam detection and feature extraction, the algorithm they proposed has good stability and accuracy. There are also welding applications in actual production. The joint tracking system that based on laser is used for multi-pass welding of thick wall in [10] and there are high production quality and production efficiency. In [11], the joint tracking system which named Smart Laser Probe and made by Meta company is used for production line of continuous welding of stainless steel.

Although there have been a lot of researches and improvements in real-time joint tracking, these systems and algorithms are used for specific situations usually and cannot be used directly in other situations, research and innovation are still needed to focus on such as optimization of the algorithms, robust and efficiency of the system.

2 Laser Welding

Laser welding is an efficient and precise welding method which using high energy density laser beam as heat source, see Fig. 2. It can realize deep penetration and narrow welds with less heat induced deformation under higher speed compared with other welding techniques e.g. spot welding [12]. It is widely used in many research areas e.g. aerospace industry, automotive industry, medical industry, etc.

The optical path and process of laser welding are shown in Fig. 2. At first, laser beams are generated in laser; then, laser beams are focused on work piece through optical focusing and pointing system which contains lens and mirrors; finally, the work piece can be welded together when the temperature reaches the melting point of the materials [13, 14].

Fig. 2 Optical path and process of laser welding

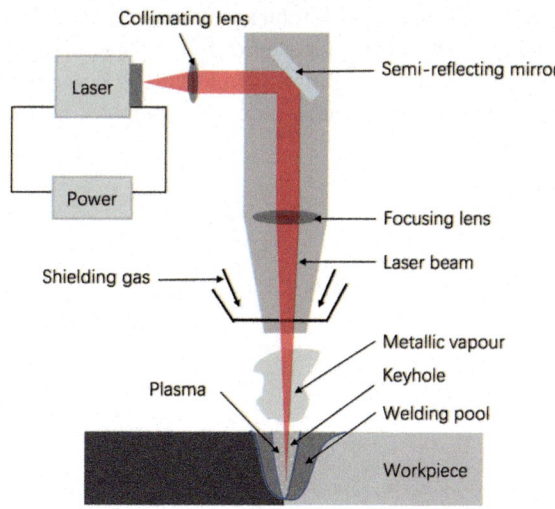

Laser welding can be clarified into heat conduction laser welding and deep penetration laser welding according to the characteristics of formation of weld joint during laser welding [12]. The welding pool of heat conduction laser welding is wide but shallow. The laser power density of heat conduction welding mode is generally in 10^4–10^5 W/cm^2, a large part of laser is reflected by the metal surface which means the light absorption rate is low. The weld depth of heat conduction model is shallow and speed is slow, it is mainly used for welding of thin work piece (thickness <1 mm) [12]. Different with heat conduction welding, deep penetration laser welding has higher power density (10^6–10^7 W/cm^2). The magnitude of the power density can cause melting and vaporization of metal materials rapidly, and a keyhole is formed in the laser irradiation point. The keyhole continues to absorb light energy and the welding pool is formed, heat diffuses to around through the welding pool. The welding pool can be affected by laser power, the larger the laser power, the deeper the welding pool will be.

3 The Real-Time Joint Tracking System

3.1 Description of a Real-Time Joint Tracking System

Joint tracking which can be seen as a form of visual serving usually focus on control-loop of sensors which focus on image acquisition, algorithms which focus on image processing, and manipulators which carry out the motion commands [15]. The composition of joint tracking system is almost the same between laser welding and other welding methods, the biggest difference is information acquisition and processing, which due to different features between different welding methods.

There is the overview of the process of a close-loop control for joint tracking, see Fig. 3. Real-time joint tracking system usually contains three parts: image acquisition part which aims to obtain the joint information usually based on cameras or other types of sensors; image processing part which focus on processing images and making decision usually based on industry computers; motion control part which carries out the commands made by computer usually includes manipulators and laser device [6].

Fig. 3 Principle of the close-loop control for joint tracking

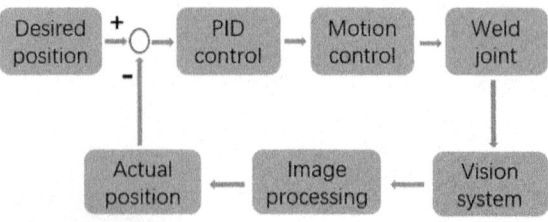

Coordinate frame

Coordinate frame is an important part that need to be considered for real-time joint tracking system. There are several coordinate frames for real-time joint tracking system: robot base coordinate frame; TCP (Tool Centre Point) coordinate frame; camera coordinate frame and work piece coordinate frame [6, 17]. In order to obtaining trajectory of robot, transformation between different coordinates is needed. Figure 4 shows frames of sensor-guided robotic laser welding.

Transformation from image plane to work piece coordinate which represents real world is needed to convert a spatial point from 2D coordinate to 3D coordinate, see Fig. 5. It can be determined by following homogenous matrix:

$$
\begin{bmatrix} u \\ v \\ 1 \end{bmatrix} = Q \begin{bmatrix} X \\ Y \\ Z \\ 1 \end{bmatrix} \tag{1}
$$

As Fig. 5 shows, X, Y, Z represent coordinate of points in workpiece frame while u, v represent coordinate of the same points in image plane. Q is a transformation matrix, which is influenced by intrinsic and extrinsic parameters of camera, can be calculated by following matrixes:

$$
K = \begin{bmatrix} \alpha_x & s & u_0 \\ 0 & \alpha_y & v_0 \\ 0 & 0 & 1 \end{bmatrix} \begin{bmatrix} R & T \end{bmatrix} \tag{2}
$$

K is matrix of intrinsic parameters of camera, which can be acquired though camera calibration. $[R\ T]$ is matrix of extrinsic parameters, R is a rotation matrix

Fig. 4 Frames of sensor-guided robotic laser welding

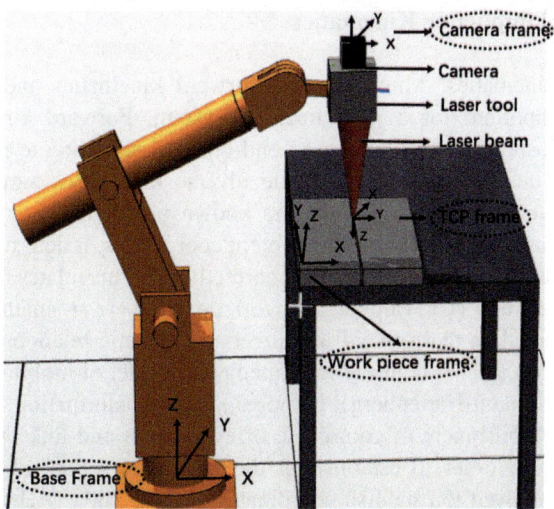

Fig. 5 Transformation from image to real world

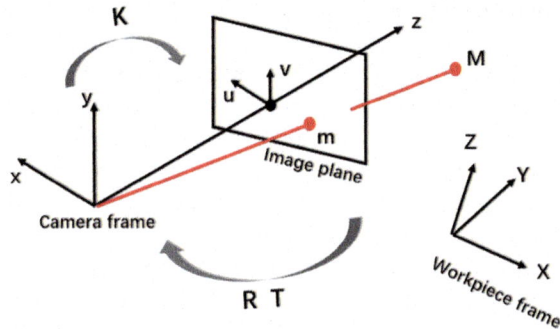

while T is a translation matrix. Points in real word can be translated into image plane according to these matrixes [6, 16].

Transformation from TCP frame to work piece frame is needed, which describes the position and orientation of TCP coordinate system with respect to work piece coordinate [6, 17]. This can be accomplished through kinematic modelling [6]. The homogenous transformation matrix can be expressed as below:

$$
{}_A^W T = \begin{bmatrix} {}_A^W R & {}_A^W P \\ 0 & 0 \end{bmatrix}
\tag{3}
$$

where A represents the coordinate of TCP while W represents the coordinate of work piece. ${}_A^W T$ is a 3×3 matrix which means transformation from TCP frame to work piece frame. ${}_A^W R$ is a 3×3 rotation matrix, which represents rotation relationship from TCP coordinate to work piece coordinate; ${}_A^W P$ is a 3×1 transfer matrix, which represents position vector that describe the position of TCP frame relative to work piece frame.

Manipulator Kinematics

Kinematics, which contains forward kinematics and inverse kinematics, is very important for a joint tracking system. Forward kinematics aims to ensure the position and orientation of end-effector according to relationship between different joints and coordinates while inverse kinematics aims to calculate the change of every joint according to the known position and orientation of end-effector and transformation between different coordinates. It determines whether the end-effector can reach the right position correctly [6]. This relates to the coordinates relationship between TCP, camera, and work piece. There are mathematic models can be used to calculate the forward and inverse kinematic relationship of joint tracking system, such as [17] built the full kinematics model of mobile welding robot by using D-H (Denavit-Hartenberg) homogeneous transformation method which is used for establishment of coordinate of every joint and link of robot.

In order to establishing the kinematic model of robot, D-H rules should be followed to establish coordinates at first. Then D-H parameters of every joint and link of robot should be measured and calculated [16, 17]. Transfer matrix that based on D-H can be expressed as follows:

$$T = \begin{bmatrix} C\theta_i & -S\theta_i & 0 & 0 \\ S\theta_i & C\theta_i & 0 & 0 \\ 0 & 0 & 1 & 0 \\ 0 & 0 & 0 & 1 \end{bmatrix} \begin{bmatrix} 1 & 0 & 0 & 0 \\ 0 & 1 & 0 & 0 \\ 0 & 0 & 1 & d_i \\ 0 & 0 & 0 & 1 \end{bmatrix} \begin{bmatrix} 1 & 0 & 0 & l_i \\ 0 & 1 & 0 & 0 \\ 0 & 0 & 1 & 0 \\ 0 & 0 & 0 & 1 \end{bmatrix} \begin{bmatrix} 1 & 0 & 0 & 0 \\ 0 & C\alpha_i & -S\alpha_i & 0 \\ 0 & S\alpha_i & C\alpha_i & 0 \\ 0 & 0 & 0 & 1 \end{bmatrix}$$

(4)

$$T = \begin{bmatrix} C\theta_i & -S\theta_i C\alpha_i & S\theta_i S\alpha_i & l_i C\theta_i \\ S\theta_i & C\theta_i C\alpha_i & -C\theta_i S\alpha_i & l_i S\theta_i \\ 0 & S\alpha_i & C\alpha_i & d_i \\ 0 & 0 & 0 & 1 \end{bmatrix}$$

(5)

$$^0_6T = {}^0_1T {}^2_2T {}^2_3T {}^3_4T {}^4_5T {}^5_6T$$

(6)

$C\theta_i = \cos\theta_i$, $S\theta_i = \sin\theta_i$, $C\alpha_i = \cos\alpha_i$, and $S\alpha_i = \sin\alpha_i$. [T] represents transfer matrix from the coordinate of one joint to another, $\begin{bmatrix} ^0_2T \end{bmatrix}$ represents the coordinate transfer from the second axis to robot base coordinate while $\begin{bmatrix} ^0_6T \end{bmatrix}$ means that the coordinate transfer from the sixth axis coordinate to robot base coordinate. Then, kinematic modelling can be established according to these equations and parameters. [17, 18]

3.2 Image Acquisition

It is very important to obtain high quality images in joint tracking system. The subsequent work will be meaningless if high quality images cannot be obtained. Acquisition of images usually influenced by e.g. the light source, the type of image sensor and disturbances like light emissions and vapour fumes during laser welding process [14].

CCD (Charge-coupled Device) and CMOS (Complementary Metal Oxide Semiconductor) are two common image sensors, which can convert optical signal into digital signal. They are both semiconductor devices, Fig. 6 shows a type of CMOS image sensor. Frame rate of camera sensor, which means frames that can be captured or transmitted per second, will influence time of image acquisition significantly, the higher the frame rate, the faster the image acquisition [13]. ROI, which means image area that selected to be acquired and processed, will affect frame rate of camera and time of image acquisition, decrease of ROI can increase frame rate and speed of image acquisition of camera sensor. ROI can be defined by users through different operators and functions.

Disturbances from laser welding process e.g. light emissions and vapour fumes will also affect image acquisition. Light emission, which mainly includes reflection

Fig. 6 A type of CMOS
image sensor

of laser beam light, heat radiation of welding pool, and radiation of laser beam, is an inevitable influence factor for image acquisition during laser welding [13]. In general, the stronger the light emission, the more difficult the image acquisition. In [19] reflected laser light can indicate weld quality, it can be influenced by the deep of the keyhole, the deeper the keyhole, the less the reflected laser light. In [20] reflected laser light is almost constant under stable welding process and will fluctuate with the state of welding conditions. It can be controlled by suitable filter. Light emission is also influenced by surface of materials, the smoother and brighter the surface, the stronger the light emission, such as there are stronger light emission for aluminum alloy compared with steel during laser welding. Light emission can be monitored by a photodiode by measuring its reverse current which related to light intensity. There is also research about vapour fume e.g. the vapour fume in CO_2 laser beam welding is monitored by a high-speed camera in [21]. As Fig. 2 shows, metallic vapour fume will be generated and diffuse over the weld joint during laser welding, metallic vapour fume that shield on weld joint is an interference for image sensor and will make image acquisition more difficult. It can be improved by different ways: e.g. extracting vapour fume by extraction system in real time; selecting suitable filter and putting in front of image sensor to reduce the influence of these disturbances.

In order to capturing images of joint in real-time, seamlessly and high-efficiency communication between camera and computer is need to be established through the interface standards which includes hardware standard and software standard [22]. Hardware interface standards aim to make sure that the hardware of vision system such as cameras, cables, and frame grabber can be connected well with each other while software standard aims to provide the same API (Application Programming Interface) for different cameras which have different hardware interfaces. There are two layers of software between camera and user application, see Fig. 7. The transport layer is the first layer which depends on the hardware interface, the main function of it is to access the camera and deliver streams. The second layer is image acquisition library which is governed by software interface standard.

Each camera has their own interface, such as the interface of CMOS camera in Fig. 6 is GigE Ethernet. SDK (Software Development Kit) which corresponds to interface of camera should be installed to build stream connection between camera

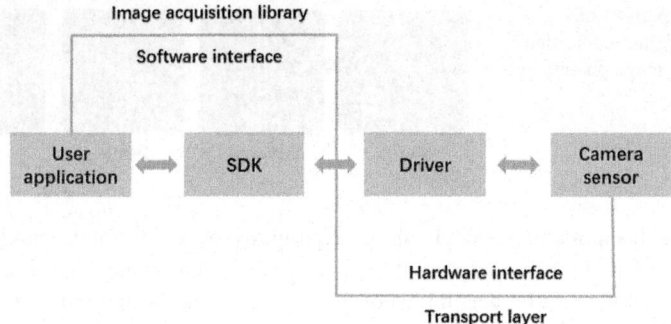

Fig. 7 Two layers of software between camera and user application

and user application through API (Application Programming Interface). In general, SDK can be found easily in Internet and there are enough programming samples which can be used directly.

3.3 Image Processing and Analysis

There is no doubt that the performance of joint tracking system will be decided by precision and speed of image processing. Image processing usually includes pre-processing, segmentation, classification, and recognition under different levels, see Fig. 8. It will be accomplished based on specific software and corresponding programming libraries and languages normally. Different programming libraries e.g. Open-CV and languages e.g. C++ can be used for image processing. Choosing of programming library and programming language should according to specific situation and requirements, because there is different performance e.g. speed or accuracy for different programming libraries and languages.

Algorithm is important for image processing and different algorithms can be used for different part of image processing. Such as in edge detection part of image processing, there are *Canny ()*, *Laplace ()*, *Sobel ()* and *Scharr ()* algorithms in Open-CV, there are different performance between them obviously, see Fig. 9.

Fig. 8 The general process of image processing

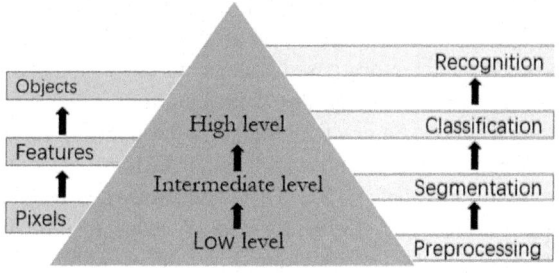

Fig. 9 Comparison of
different algorithms for edge
detection of image processing
of butt joint

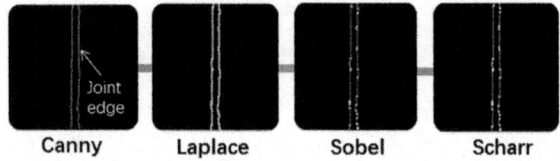

There have been many research about algorithms used for joint tracking. The modified Hough algorithm is used in [23] to process joint images, the results show that seam tracking can be accomplished in a shorter time in that paper. Reference [24] introduced a simple seam-tracking algorithm through characteristic-point detection using a laser-displacement sensor to detect the seam of single-butt welding with manually tack-welded non-zero gaps. Reference [25] used Kalman filtering algorithm to eliminate the influence of image noise and reduce error between the measurement position and real position of the objects. Improvement of the algorithms can promote the development of the welding, such as in threshold selection part of image processing for joint tracking, adaptive algorithm for thresholding is faster than manually selecting of threshold, which increases the react efficiency and improves the welding accuracy.

Algorithms are always the most important part for image processing, good algorithms should have higher precision and faster speed which may be the bottleneck of image processing. Although there have been many algorithms to use, these algorithms are not effective for all the conditions, so, these algorithms cannot be used directly usually. Improve, optimize and even create new algorithms according to the specific situation is needed.

3.4 Motion Control

After getting actual position of joint through image processing and analysis, decision should be made and sent to motion control part which includes control of robotic motion and welding parameters. Robot will receive signals from its controller and make corresponding compensation action, and then feedback the real-time position to computer in closed loop control system, see Fig. 3. The deviation should be compensated and get good welding quality in theory after implementation of motion control, see Fig. 10.

There has been much research about robot motion control system e.g. [14] proposed a trajectory-based control system which can generate trajectory of robot in real-time according to the data acquired from sensors. However, there is only limited success in practical industrial production. Externally guided motion (EGM), as a part of Robotware which is an ABB robot control software, can be used for robot motion control [26]. Tested accuracy and feasibility of EGM in laser welding application under different types of paths, results show that there is only minor influence which can be ignored in laser welding application.

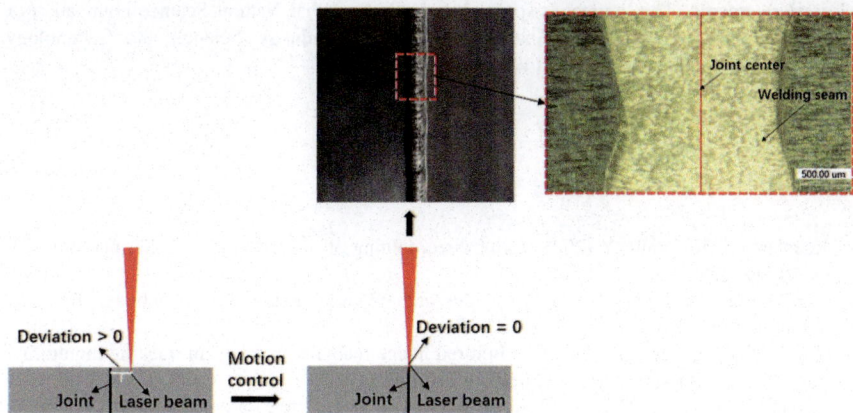

Fig. 10 Ideal motion control lead to zero deviation between laser beam port and joint center

In order to accomplishing motion control, mathematical model should be established to convert deviation signal got from image process and analysis part to signal of direction, distance or angle of robot joints in real time. The mathematical model can be built based on different software such as MATLAB.

4 Peroration

Considering limitations of robotic laser welding, which induces by real time changes in joint or environment, a real-time joint tracking system is needed to compensate real time changes and keep the laser beam following joint center consistently. In this paper, laser welding and composition of a real-time joint tracking system which mainly includes image acquisition part, image process and analysis part, and motion control part is introduced. Frames and kinematics modeling of joint tracking in robotic laser welding is also presented. Each part of joint tracking system can be influenced by different factors, such as image process and analysis part can be influenced by algorithms, programming library and programming language, etc. Communication between different part of joint tracking system can also influence the performance of the whole system through interface, API, etc.

Many relevant investigations of real-time joint tracking are reviewed in this paper, many systems and algorithms are proposed and successfully used for real-time joint tracking. However, these systems and algorithms can only be used in limited situations which mean specific applications, so, research and innovation are still needed to focus on improvement of robust and efficiency of the system, commercial solutions for joint tracking, etc.

Acknowledgements This project is sponsored by the Shanghai Natural Science Foundation of China (14ZR1418800), and the Shanghai Automotive Industry Science and Technology Development Foundation of China (1404).

References

1. Katayama S, Kawahito Y (2008) Laser direct joining of metal and plastic. Scripta Mater 59 (12):1247–1250
2. Wu Q, Gong JK, Chen GY et al (2008) Research on laser welding of vehicle body. Opt Laser Technol 40(2):420–426
3. Gao X, You D, Katayama S (2012) Infrared image recognition for seam tracking monitoring during fibre laser welding. Mechatronics 22(4):370–380
4. Huang W, Kovacevic R (2012) Development of a real-time laser-based machine vision system to monitor and control welding processes. Int J Adv Manuf Technol 63(1–4):235–248
5. Li Y et al (2006) Girth seam tracking system based on vision for pipe welding robot. In: Proceedings of 2006 international conference on robotic welding, intelligence and automation, vol 362, Springer, Shanghai, pp 391–399
6. Nayak N, Ray A (2013) Intelligent joint tracking for robotic welding. Nayak N & Ray A, London, pp 1350–1351
7. Huang Y, Xiao Y, Wang P et al (2013) A seam-tracking laser welding platform with 3D and 2D visual information fusion vision sensor system. Int J Adv Manuf Technol 67(1–4): 415–426
8. Gu WP, Xiong ZY, Wan W (2013) Autonomous seam acquisition and tracking system for multi-pass welding based on vision sensor. Int J Adv Manuf Technol 69(1):451–460
9. Wang XP et al (2014) Weld joint detection and feature extraction based on laser vision. In: Proceedings of 33rd Chinese control conference, IEEE, Nanjing, pp 8249–8252
10. Lin SB, Yang CL, Beattie RJ (2004) Applications of laser seam tracking to welding thick wall vessels. Weld Joining 11:15–18
11. Shen JM, Du XJ, Lu MH (2012) Application of smart laser probe to continuous welded stainless pipe line. Technol Dev Enterp 31(25):34–35
12. Wang ZH (2011) Welding process. Beijing normal university publishing group, Beijing, pp 188–194
13. Nilsen M (2017) Optical detection of joint position in zero gap laser beam welding. Dissertation, University West
14. Steen WM, Mazumder J (2010) Laser material processing. Springer Science & Business Media, London, pp 157–199
15. Agapakis JE (1990) Approaches for recognition and interpretation of workpiece surface features using structured lighting. Int J Robot Res 9(5):3–16
16. Fan OY et al (2012) Offline kinematics analysis and path planning of two-robot coordination in exhaust manifold welding. In: Proceedings of international conference on robotics and biomimetics, IEEE, Guangzhou, pp 1806–1811. https://doi.org/10.1109/ROBIO.2012.6491230
17. Verma DA, Gor M (2010) Forward kinematics analysis of 6-dof arc welding robot. Int J Eng Sci Technol 2(9):4682–4686
18. Wang ZX, Fan WX, Zhang BC et al (2012) Kinematical analysis and simulation of industrial robot based on Matlab. J Mech Electr Eng 29(1):34–37
19. Huegel H et al (1999) Laser beam welding: recent developments on process conduction and quality assurance. In: Proceedings of tenth international school on quantum electronics: laser physics and applications, vol. 3571, SPIE, Germany, pp 52–60
20. Eriksson I, Kaplan A (2009) Evaluation of laser weld monitoring: A case study. In: Proceedings of ICALEO, vol. 102, DiVA, Sweden, pp 1419–1425

21. Cai Y, Yang Q, Sun D et al (2014) Monitoring of deviation status of incident laser beam during CO_2 laser welding processes for I-core sandwich construction. Int J Adv Manuf Technol 77(1–4):305–320

22. Global machine vision interface standards (2014) Standards for machine vision interface. https://www.visiononline.org/vision-standards.cfm.

23. Wu QQ, Lee JP, Park MH et al (2015) A study on the modified Hough algorithm for image processing in weld seam tracking. J Mech Sci Technol 29(11):4859–4865

24. Chang D, Son D, Lee J et al (2012) A new seam-tracking algorithm through characteristic-point detection for a portable welding robot. Robot Comput Integr Manuf 28(1):1–13

25. Gao XD, Na SJ (2005) Detection of weld position and seam tracking based on Kalman filtering of weld pool images. J Manuf Syst 24(1):1–12

26. Gao JM (2016) Industrial robot motion control for joint tracking in laser welding. Dissertation, University West

Effects of Process Parameters on the Weld Quality During Double-Pulsed Gas Metal Arc Welding of 2205 Duplex Stainless Steel

Yu Hu and Jiaxiang Xue

Abstract In this paper, double-pulsed melting inert-gas welding (MIG) was conducted on 2205 duplex stainless steel and the effects of the number of strong pulses, the number of weak pulses and the welding speed were studied. The electrical parameters in the welding process were collected through the wavelet analyzer. The waveform charts of the current and voltage, the U-I diagram, the energy input, the dynamic resistance and other real-time signals were analyzed. In addition, mechanical tensile and metallographic tests were conducted. The results demonstrated that the welding speed had the highest impact on the welding quality among all the three factors, followed by the number of weak pulses and the number of strong pulses. The welded seam obtained by various numbers of strong and weak pulses was relatively uniform, indicating that the effect of the number of strong and weak pulses on the welded seam was relatively low. The tensile fracture occurred in the base material part of the duplex stainless steel, indicating that the welded seam region had a higher tensile strength than the base material. The tensile strengths of various specimens were similar. When the welding speed was reduced, the heat input increased during the welding of the welded seam, the cooling rate of the welded seam slowed down, whereas the ferrite transition duration increased, which led to the formation of higher amount of austenite and a relatively coarse structure.

Keywords Duplex stainless steel · Double pulse · MIG · Welding quality

1 Introduction

Duplex Stainless Steels (DSS) combine the excellent toughness and weldability of austenitic stainless steels and the high strength and resistance against the oxide stress corrosion of the ferritic stainless steel. Therefore, the DSS display the merits

Y. Hu · J. Xue (✉)
School of Mechanical and Automotive Engineering, South China University of Technology, Guangzhou 510641, China
e-mail: mejiaxue@scut.edu.cn

© Springer Nature Singapore Pte Ltd. 2018   113
S. Chen et al. (eds.), *Transactions on Intelligent Welding Manufacturing*,
Transactions on Intelligent Welding Manufacturing,
https://doi.org/10.1007/978-981-10-8330-3_7

of good resistance against pitting, crevice corrosion, stress corrosion and corrosion fatigue, exhibiting corrosion resistance, good comprehensive mechanical properties and other characteristics. Therefore, the application of DSS has rapidly been developed and it is widely utilized in petroleum, chemical, marine and power industries [1, 2].

Although the TIG welding can acquire the fish-scale appearance, the corresponding production efficiency is low. Consequently, it is difficult to meet the requirements of large-scale production. The pulse MIG welding has higher production efficiency and is easy for automated production to be achieved, whereas certain issues exist in the welding quality [3, 4]. The current significantly advanced welding methods include the pulsed melting inert gas welding (P-MIG), the double wire melting argon arc welding, the alternating current pulsed melting inert gas welding (AC-PMIG), the double-pulsed gas metal arc welding (DP-GMAW) and the friction stir welding. The MIG welding process through the double pulse is the utilization of a low-frequency signal to modulate the high-frequency pulse current, in order for the current waveform chart the present an alternative appearance of the strong pulse and weak pulse groups. Such a welding method not only has the characteristics of single-pulse welding such as the low average input current, whereas it also stirs the molten pool and releases the bubbles, resulting in the fish-scale weld seam. At present, the domestic and international studies on the double-pulsed MIG mainly are focused on the arc behavior, the welding speed, the joint performance and the post-treatment subsequently to welding [5]. In contrast, the researches on the effects of the number of strong and weak pulses and the welding speed on the welding process for stainless steels are rarely reported. In order to understand the effects of the number of both double pulse and welding speed on the welded seam shaping, the MIG welding was performed on a 2205 duplex stainless steel through various numbers of strong and weak pulses. The pulse number effects on the formation, the welding stability and the mechanical properties and microstructure of the welded seam were investigated, providing reference to the processing parameters selection in the double-pulsed welding of duplex stainless steels [6, 7].

2 Test Methods

2.1 Test Material

The test material utilized in the test was a 4.0 mm-thick duplex stainless-steel plate. The ER2209 welding wire with of 1.2 mm in diameter was selected. The chemical compositions of these two materials are presented in Table 1.

Table 1 Chemical composition of 2205 and ER2209

Element	C	Mn	Si	Cr	Ni	Mo	N	S	P
2205	0.024	1.4	0.62	21.3	5.4	3.0	0.16	0.01	0.024
ER2209	0.017	1.73	0.52	22.43	8.45	2.94	0.15	0.008	0.012

2.2 Test Platform

The data acquisition platform for the welding test included a load resistance box, an industrial personal computer (Advantech 610), a data acquisition card (Advantech PCL 1800), a Huawei automatic walking control device, an arc dynamic wavelet analyzer for the real-time signals acquisition of the voltage and current, a soft switching inverter based on the digital signal processing (DSP) for pulsed MIG welding, a wire feeding machine and other devices.

2.3 Process and Parameters

The self-developed digital multi-functional power supply was adopted in the test. A welding arc dynamic wavelet analyzer was utilized to measure the voltage and the current. The arc dynamic wavelet analyzer was an important tool to collect and analyze the real-time signal during welding. The corresponding host computer software was installed on the industrial personal computer to collect the real-time current and voltage signals during welding. Following the analysis and processing, the signals could be utilized to determine the welding stability. Prior to welding, the oxide film on the base metal surface was polished with the No. 800 grit sandpaper, whereas the grease residue on the base metal surface was cleaned with acetone.

An orthogonal test table was designed based on the number of strong pulses, the number of weak pulses and the welding speed, as presented in Table 2. Respectively, the numbers of strong pulses were 10, 12, 14 and 16; the numbers of weak pulses were 5, 6, 7 and 8; the welding speeds were 0.6, 0.8, 1.0 and 1.2 m/min. The peak and base current of the strong pulse group were 440 and 110 A; the peak and base current of the weak pulse group were 280 and 70 A. The protective gas was argon (Ar) with the purity of 99.99% and the flow rate of 18 L/min. The welding object was the duplex stainless-steel plate with the thickness of 4 mm and the surfacing welding was executed.

Table 2 Orthogonal test table

No.	No. of strong pulses N_1	No. of weak pulses N_2	Welding speeds (m/min)
1	10	5	0.6
2	10	6	0.8
3	10	7	1.0
4	10	8	1.2
5	12	5	0.8
6	12	6	0.6
7	12	7	1.2
8	12	8	1.0
9	14	5	1.0
10	14	6	1.2
11	14	7	0.6
12	14	8	0.8
13	16	5	1.2
14	16	6	1.0
15	16	7	0.8
16	16	8	0.6

3   Test Result and Analysis

3.1   Appearance of Welded Seam

According to the welding quality and uniformity of the welded seam, the comprehensive evaluation indicator, A, and three subordinate evaluation indicators: A1 (welding stability), A2 (with or without arc breaking) and A3 (with or without splash) were determined. According to the fuzzy comprehensive evaluation model, the synthesis product with fuzzy relation and the matrix could be obtained as: $S = A*R\cdot F$ [8], where A is the indicator weight as estimated by experts, $A = (A1, A2, A3) = (0.8, 0.15, 0.05)$. According to the principle of difference scoring method, the Vij ($j = 1, 2, 3$) was differentially scored. The evaluation set was $F = (f1, f2, f3)T = (100, 75, 50)T$. The evaluation on every test indicator is carried out by the group composed of five experts. As an example, the comprehensive evaluation matrix for the index A1 of weld seam in test 1 was $V1 = (V11, V12, V13) = (0.4, 0.2, 0.4)$, the $V2$ and $V3$ can be obtained according to the same principle. The comprehensive score was calculated as $S1 = A1*R1\cdot F = 76$. Similarly, the other test scores could be obtained. The evaluation table of the experts is presented as Table 3.

All tests, based on the parameters in Table 2, could be successfully completed and the formed welded seams were continuous and uniform. The undercut and any other defects were seldom observed. With reference to the orthogonal test table, it could be observed from the seam images that the welded surface formed in test 1

Table 3 Evaluation table given by experts

Stability A_1			Arc breaking A_2			Splash A_3			Comprehensive score for A
Stable	Common	Poor	No	Common	Obvious	No	Common	Obvious	
V_{11}	V_{12}	V_{13}	V_{21}	V_{22}	V_{23}	V_{31}	V_{32}	V_{33}	
2	1	2	3	1	1	2	2	1	76
2	1	2	2	1	2	2	1	2	72
2	2	1	3	1	1	2	2	1	82
2	1	2	1	2	2	3	0	2	74
2	2	1	2	2	1	3	1	1	78
2	1	2	2	2	1	2	2	1	75
1	3	1	2	2	1	2	1	2	74
4	1	0	2	2	1	3	1	1	80
4	1	0	3	1	1	3	1	1	85
1	2	2	2	1	2	3	1	1	73
3	0	2	1	2	2	3	1	1	81
3	1	1	3	1	1	1	2	2	78
1	1	3	1	1	3	2	0	3	72
1	2	2	1	0	4	2	1	2	71
2	2	1	2	2	1	1	3	1	76
3	1	1	2	2	1	3	1	1	77

was smooth; without an apparent fish-scale pattern; the edge of the weld was not significantly flat. Therefore, the score given by the evaluation system was approximately 76 points. The welding in test 9 was quite stable, with less splashing and displayed the best quality of the weld. The two sides of the welded seam were flat and uniform, whereas a fish-scale weld groove could be observed. The splashing in welding process 2 was less, whereas the welded seam width was inconsistent. An apparent snake-shaped pattern appeared and the fish-scale pattern was not observed. Following the orthogonal test data processing (Table 4), it could be concluded that among the number of strong pulses, the number of weak pulses and the welding speed, the welding speed had the highest impact on the weldment quality. The weldments are presented in Fig. 1.

3.2 Analysis on Real-Time Data Acquired During Welding

Waveform chart of current

The test 9 with the improved welding quality was selected as the real-time data analysis object. The real-time current signal acquired during welding is presented in Fig. 2. Figure 2a, b are the waveform charts of the current. The current waveform in the entire welding process was well reproducible and no instantaneous burr was

Table 4 Orthogonal test results

Test No.	Number of strong pulses N_1	Number of weak pulses N_2	Welding speeds (m/min)	Comprehensive score
1	10	5	0.6	76
2	10	6	0.8	72
3	10	7	1.0	82
4	10	8	1.2	74
5	12	5	0.8	78
6	12	6	0.6	75
7	12	7	1.2	74
8	12	8	1.0	80
9	14	5	1.0	85
10	14	6	1.2	73
11	14	7	0.6	81
12	14	8	0.8	78
13	16	5	1.2	72
14	16	6	1.0	71
15	16	7	0.8	76
16	16	8	0.6	77
Average k_1	76.00	77.75	77.25	
Average k_1	76.75	72.75	76.0	
Average k_1	79.25	78.25	79.5	
Average k_1	74.0	77.25	73.25	
Range, R	5.25	5.5	6.25	
Order	Welding speed > Number of weak pulses > Number of strong pulses			

present. The statistical average current was 200 A; the base current of the strong pulse group was 110 A and the peak current was 440 A; the peak current of the weak pulse group was 280 A and the base current was 70 A.

Voltage waveform chart

The arc length stability in the welding process directly represents the voltage stability. Figure 3a presents the waveform chart of the welding voltage. The statistical average voltage was 26.27 V, with the minimum voltage of 17.79 V and the maximum voltage of 38.66 V. The entire waveform appeared to change regularly. The lower oscillation range of the voltage indicated a stable welding. Figure 3b presents the amplified waveform of the voltage. It could be observed that the voltage became higher subsequently formed a peak when the current approached the corresponding peak, which corresponded to the arc length sudden elongation

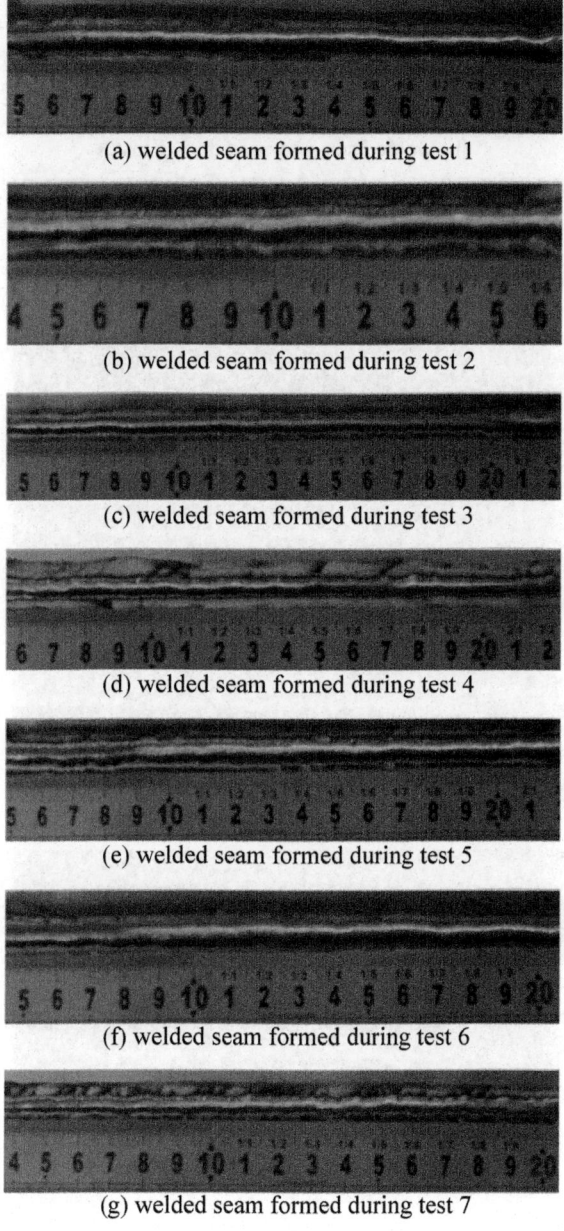

(a) welded seam formed during test 1

(b) welded seam formed during test 2

(c) welded seam formed during test 3

(d) welded seam formed during test 4

(e) welded seam formed during test 5

(f) welded seam formed during test 6

(g) welded seam formed during test 7

Fig. 1 Welded seams formed during orthogonal test

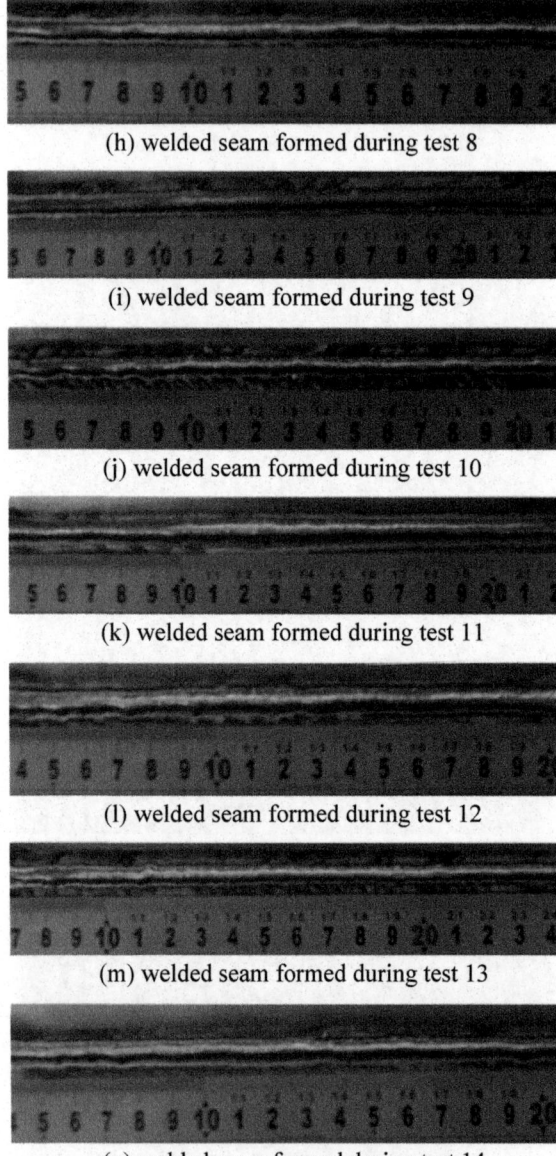

(h) welded seam formed during test 8

(i) welded seam formed during test 9

(j) welded seam formed during test 10

(k) welded seam formed during test 11

(l) welded seam formed during test 12

(m) welded seam formed during test 13

(n) welded seam formed during test 14

Fig. 1 (continued)

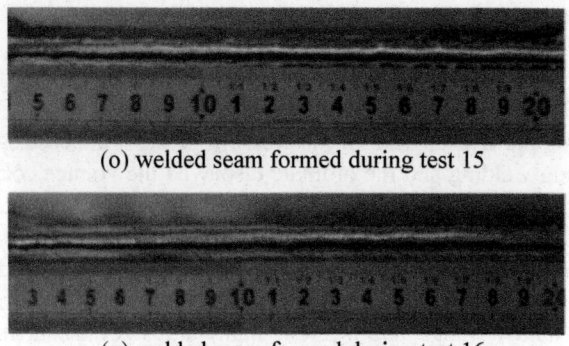

(o) welded seam formed during test 15

(p) welded seam formed during test 16

Fig. 1 (continued)

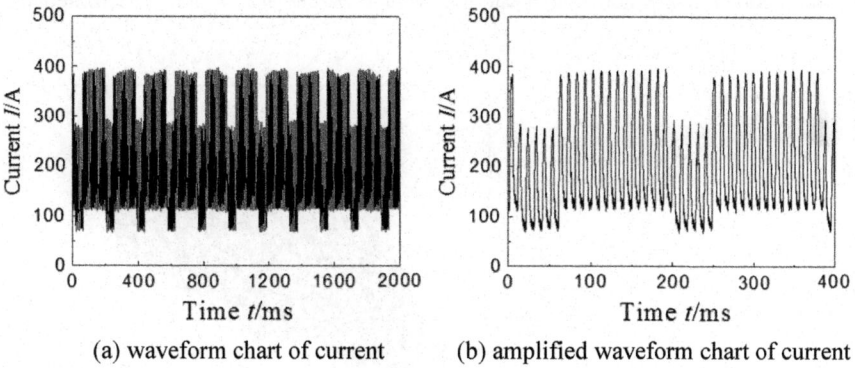

(a) waveform chart of current   (b) amplified waveform chart of current

Fig. 2 Real-time signal of current during welding

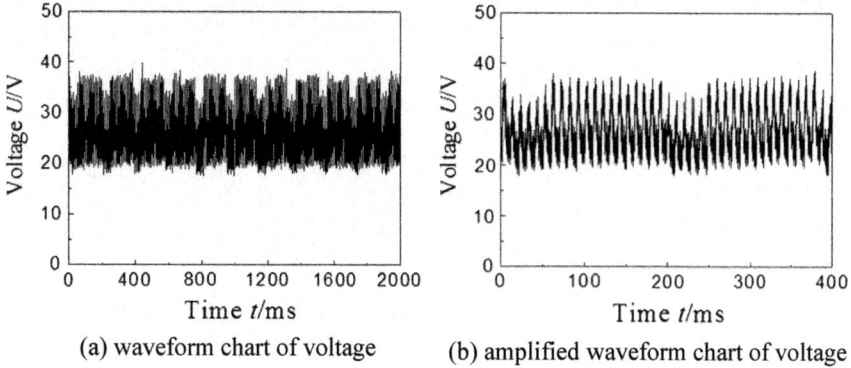

(a) waveform chart of voltage   (b) amplified waveform chart of voltage

Fig. 3 Real-time voltage signal during welding

following the melt droplet transition. In contrast, in the pilot arc base current stage, the voltage was relatively low and stable, with lower oscillations.

U-I diagram

Figure 4 presents the U-I diagram of the welding process, the abscissa represented the current during welding and the ordinate displayed the voltage corresponding to the current, as a combination of Figs. 2 and 3. From the U-I diagram the dynamic welding could be analyzed and evaluated, intuitively. The edge lines drawn by the current and the voltage were clear and neat, along with centralized distribution, which indicated that the current and the voltage during welding were concentrated in a relatively narrow range and without high jumps. Consequently, the entire welding displayed a good stability.

Input energy of welding and dynamic resistance

The input energy of the welding, presented in Fig. 5, is the product of the instantaneous current and instantaneous voltage of the arc, which determined the

Fig. 4 U-I diagram of welding

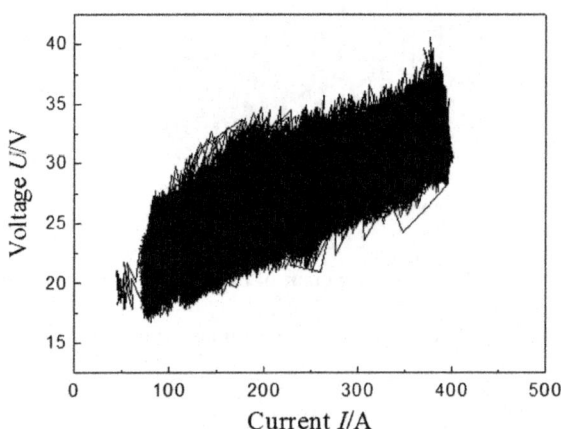

Fig. 5 Input energy of welding process

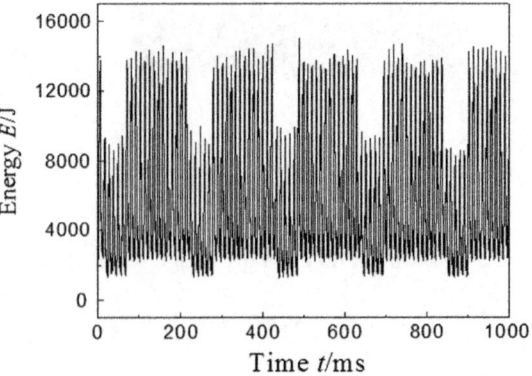

Fig. 6 Dynamic resistance

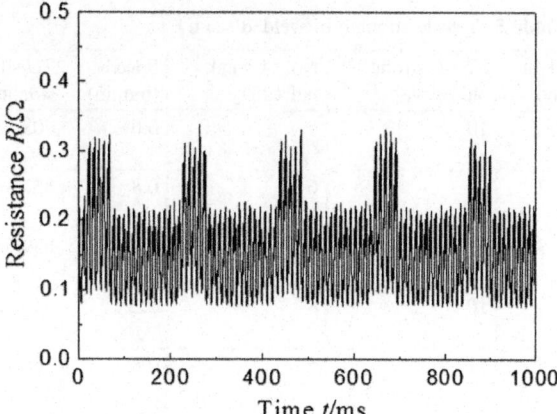

arc length and the transition form of the droplet [9–12]. As it could be observed, the pulse energy waveform displayed a regular rectangular change, indicating that the droplet transition during welding was regular. Figure 6 presents the dynamic resistance during welding, which was the ratio of the instantaneous voltage and current. It could be observed that the resistance fluctuated within the 0.1–0.3 Ω range around the 0.2 Ω, indicating a good concentration and regularity. From the welding tests, the entire welding was successfully complete and no arc breaking or short circuit phenomena were observed.

3.3   Analysis on Mechanical Properties of Welded Seam

The surfaces of the specimen following welding were polished and the tensile specimens were prepared through wire cutting. The thickness of the specimens was 4 mm. The tensile test result of the welded seams is presented in Table 5. The fracture position occurred in base material part of the duplex stainless steel, indicating that the welded seam had a higher tensile strength than the base material. The difference in the tensile strength of various samples was not significantly high, which was consistent with the aforementioned conclusion that the fracture occurred in the base metal part rather than the fusion and the heat-affected zones.

3.4   Metallographic Analysis of Welded Seam

The weld seams in test 9 had improved welding quality and were consequently selected as the analysis objects in the metallographic analysis. The welded seam could be divided into three zones: the molten pool zone, the fusion zone and the heat-affected zone. The microstructures of the fusion zone observed with the

Table 5 Tensile strength of welded seam

Test no.	No. of strong pulses N_1	No. of weak pulses N_2	Speed (m/min)	Tensile strength (MPa)	Fracture position
1	10	5	0.6	903.12	Lower region of base material
2	10	6	0.8	857.3	Upper region of base material
3	10	7	1.0	859.66	Lower region of base material
4	10	8	1.2	850.42	Upper region of base material
5	12	5	0.8	868.47	Upper region of base material
6	12	6	0.6	873.91	Lower region of base material
7	12	7	1.2	863.31	Upper region of base material
8	12	8	1.0	865.43	Lower region of base material
9	14	5	1.0	861.84	Lower region of base material
10	14	6	1.2	863.15	Lower region of base material
11	14	7	0.6	889.21	Lower region of base material
12	14	8	0.8	870.04	Lower region of base material
13	16	5	1.2	873.86	Lower region of base material
14	16	6	1.0	872.42	Lower region of base material
15	16	7	0.8	887.05	Lower region of base material
16	16	8	0.6	904.86	Upper region of base material

50-fold and 500-fold microscope are presented in Figs. 7 and 8, respectively. The microstructure had an irregular strip feature and an alternating distribution of two phases, such as the substrate was ferrite whereas the strip, lump and feathery organizations were austenite. The molten pool zone, presented in Fig. 9, demonstrated the as-cast dendritic crystal structure that was mainly composed of the austenite structure. The precipitates of high-sized tracts of feathery and dendritic austenite were observed, indicating that the metal in the molten pool had poorer hardness and toughness, along with increased brittleness. The microstructure of the base metal is presented in Fig. 10. The structure had a clear orientation and a

Fig. 7 Metallographic image of fusion zone (50 ×)

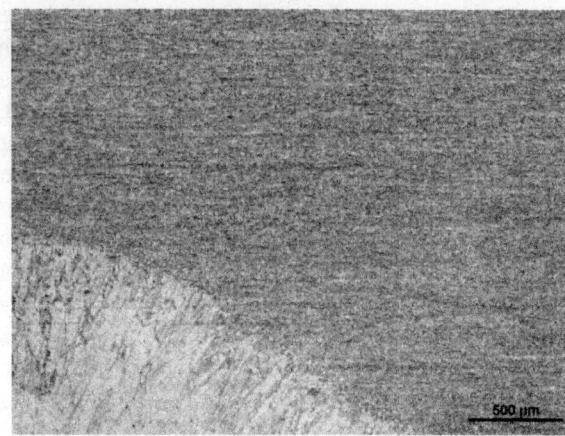

Fig. 8 Metallographic image of fusion zone (500 ×)

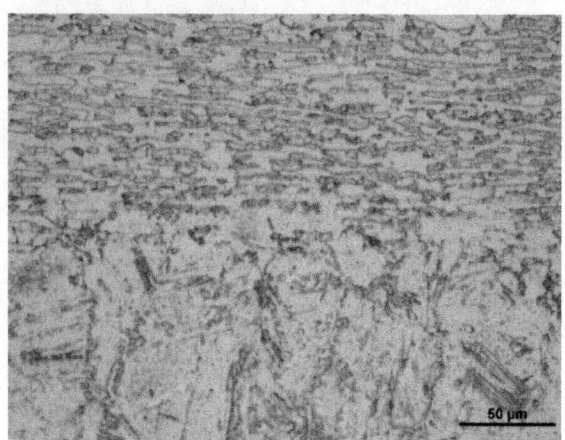

Fig. 9 Metallographic image of molten pool (500 ×)

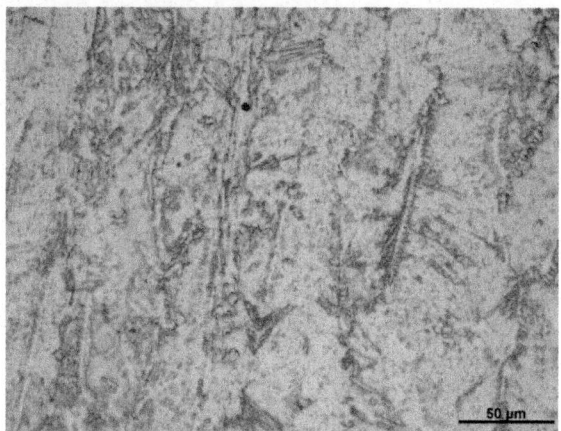

strip-like distribution. The strip-like austenite distributed on the ferrite substrate and the proportions of the two components were similar. The darker part was the ferrite, whereas the lighter part was the austenite. The microstructure of the heat-affected zone is presented in Fig. 11, where the grains were not significantly coarsened, still belonging to the typical dual-phase morphology of the austenite and ferrite.

The metallographic graphs of the molten pool zones in tests 1, 6, 9, and 13 are presented in Figs. 12, 13, 9 and 14, whereas the corresponding welding speeds were 0.6, 0.8, 1.0 and 1.2 m/min, respectively. As it could be observed from these figures, as the welding speed decreased, the heat input during welding increased and the cooling rate of the welded seam became slower, increasing the transition duration and promoting the formation of austenite, resulting in a relatively coarse structure.

Fig. 10 Metallographic image of base material (500 ×)

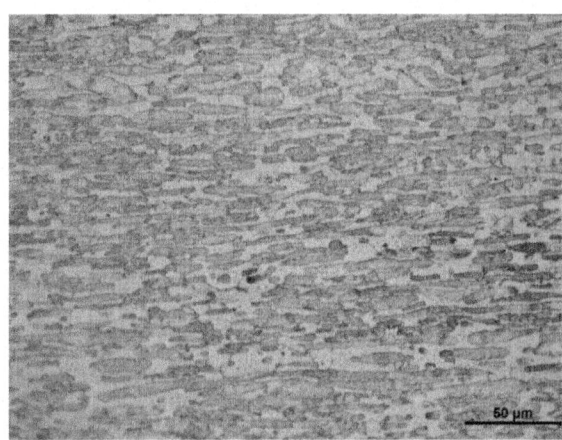

Fig. 11 Metallographic image of HAZ (500 ×)

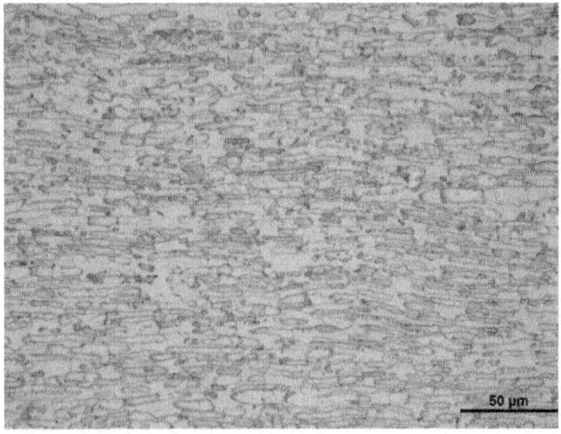

Fig. 12 Metallographic
image of molten pool (500 ×)

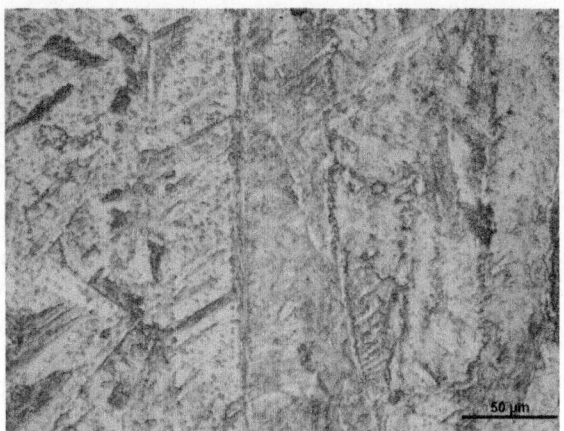

Fig. 13 Metallographic
image of molten pool (500 ×)

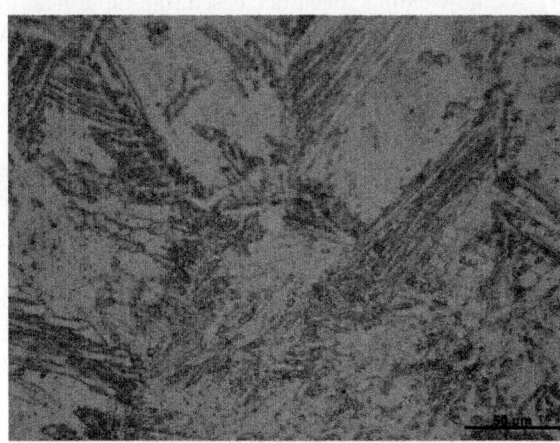

Fig. 14 Metallographic
image of molten pool (500 ×)

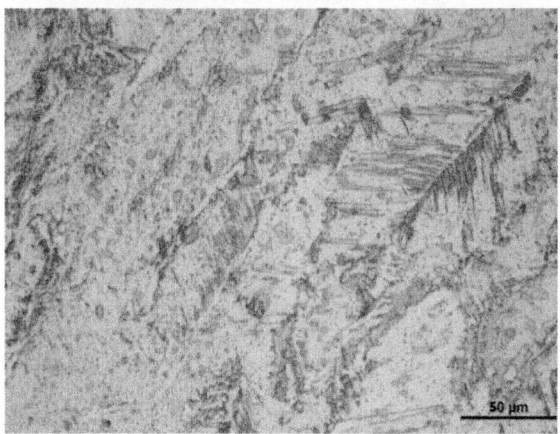

4   Conclusion

(1) During the welding tests of the 4 mm-thick duplex stainless-steel plate, the welding was smooth and the splashes were less. The welded seam was good in appearance, displaying uniform and bright fish-scale morphology. The arc voice was soft and no melting penetration phenomenon was observed.

(2) Among the aforementioned three factors, the order in affecting extent was weld speed > number of weak pulses > number of strong pulses. The welded seams obtained under various numbers of strong and weak pulses were relatively uniform, indicating that the number of strong and weak pulses only had a slight effect on the welded seam.

(3) The tensile fracture occurred in the base material part of the duplex stainless steel, indicating that the welded seam had a higher tensile strength than the base metal, whereas the difference in the sample tensile strengths was insignificant.

(4) As the welding speed decreased, the input heat during welding increased and the cooling rate of the welded seam became slower, increasing the transition duration of ferrite and promoting the formation of higher amount of austenite, resulting in a relatively coarse structure.

Acknowledgements The research was sponsored by the High-level Leading Talent Introduction Program of GDAS (2016GDASRC-0106), the Natural Science Foundation of Guangdong (2016A030313117), the Industry-university-research of Guangdong province and Ministry of Education (2013B090600098).

References

1. Wang Z, Xu H, Wu W et al (2011) Joint performance of duplex stainless 2205 by laser MIG hybrid welding. Trans China Weld Inst 32(2):105–108
2. Wang Z, Han J, Song H et al (2011) Comparative analysis for joint performance of duplex stainless steel by different arc welding methods. Trans China Weld Inst 32(4):37–40
3. Yao P, Jiaxiang Xue, Qiang Zhu et al (2014) Quantitative evaluation of double wire pulsed welding stability based on probability density distribution. Trans China Weld Inst 35(7):51–54
4. Yao P, Xue J, Zhong L et al (2012) Intelligent process expert database of double pulse MIG welding of Al-Si alloy. Trans China Weld Inst 21(1):59–63
5. Xie H, Wang L, Sha X et al (2015) Analysis of strong and weak pulse ratio on weld quality of AA6061 aluminum alloy double pulsed gas metal arc welding. Trans China Weld Inst 36(12):77–80
6. Yoganandh J, Kannan T, Kumaresh Babu SP et al (2013) Optimization of GMAW process parameters in austenitic stainless-steel cladding using genetic algorithm based computational models. Exp Tech 37(5):48–58
7. Chen H, Xue J, Heng G (2016) Improvement of double wire mig welding by using sine wave pulse modulation control method. In: IHMSC-2016 eighth international conference on intelligent human-machine systems and cybernetics, pp 451–455
8. Zhu Q, Xue J, Xu M (2015) A probe into backward median current waveform welding of aluminum alloys. J S China Univ Technol (Natural Science Edition) 43(3):15–20

9. Zhu Q, Xue J, Xu M (2016) Gaussian pulsed MIG welding of aluminum alloy sheet. Trans China Weld Inst 37(8):71–74
10. Niu Y, Xue H, Li H (2012) Effect of peak pulse voltage on metal transfer and formation of weld in double-wire pulsed MIG welding. Trans China Weld Inst 31(1):50–54
11. Aktepe A, Ersöz S, Lüy M (2014) Welding process optimization with artificial neural network applications. Neutral Netw World 24(6):655–670
12. Wang LL, Wei HL, Xue JX et al (2017) A pathway to microstructural refinement through double pulsed gas metal arc welding. Scripta Mater 134:61–65

Grain Boundary Feature and Its Effect on Mechanical Property of Ni 690 Alloy Layer Produced by GTAW

Wangteng Lin, Xiao Wei, Shaofeng Yang, Yuqian Huang,
Maolong Zhang, Weihua Liu, Jijin Xu, Junmei Chen,
Chun Yu and Hao Lu

Abstract Ni690 alloy surfacing layers were fabricated by gas tungsten arc welding (GTAW) with two different heat inputs, namely, large heat input (LHI) and small heat input (SHI). The high temperature performance of the surfacing layer was evaluated by employing Gleeble 3500 thermal/mechanical simulator. It is found that the ultimate tensile strength (UTS) of the LHI samples was higher than that of the SHI samples after reheat thermal cycles, regardless of the reheating temperature. The EBSD result shows that the proportion of high angle grain boundaries (GBs, >15°) in the LHI sample was obviously higher than that in the SHI sample. And more $M_{23}C_6$ particles were found to precipitate at the high angle GBs. The relations among UTS, GB angle distribution and $M_{23}C_6$ precipitations were analyzed. Moreover, the fracture modes were characterized by optical microscope (OM) and scanning electron microscope (SEM). The fracture mode was ductile fracture with deep dimples at 700 °C. While it changed to brittle intergranular fracture at 900 °C. As the temperature was enhanced to 1050 °C, the fracture returned to transgranular mode, with shallow dimples.

Keywords Nickel based alloy · Microstructure · Grain boundary precipitates Mechanical property · Welding

W. Lin · X. Wei · S. Yang · Y. Huang · M. Zhang · W. Liu
J. Xu · J. Chen · C. Yu (✉) · H. Lu (✉)
Key Lab of Shanghai Laser Manufacturing and Materials Modification, School of Materials Science and Engineering, Shanghai Jiao Tong University, Shanghai 200240, China
e-mail: yuchun1980@sjtu.edu.cn

H. Lu
e-mail: shweld@sjtu.edu.cn

M. Zhang
Shanghai Electric Nuclear Power Equipment Co., Ltd., Shanghai 201306, China

W. Liu
China Nuclear Industry Fifth Construction Co., Ltd., Shanghai 201512, China

© Springer Nature Singapore Pte Ltd. 2018
S. Chen et al. (eds.), *Transactions on Intelligent Welding Manufacturing*,
Transactions on Intelligent Welding Manufacturing,
https://doi.org/10.1007/978-981-10-8330-3_8

1 Introduction

Ni-Cr–Fe alloys are widely used in manufacturing of nuclear power equipment (like steam generator tubes), because of its excellent corrosion resistance and high temperature performance [1]. 600 alloy, the content of Cr was 15%, was once selected as one of the nuclear power equipment materials [2], but it is easy to produce stress corrosion cracking (SCC), which is related to the poor Cr phenomenon near the grain boundaries (GBs). Then, 690 alloy (the content of Cr was up to 30%) was developed as a superior substitute material in pressurized water reactors (PWR) to avoid the SCC [3]. In practical engineering, multi-pass welding technology is often used, which is an essential process to fabricate the nuclear power application [4]. Lin [5] employed GTAW to study the relationships between microstructure and properties of buffer layer with Inconel 52M clad on AISI 316L stainless steel. And Lee and Chen [6] found that the phase transformations caused by the heat input in the traditional gas tungsten arc welding (GTAW) welding procedure. The heat affected zone (HAZ) should be subjected to at least two thermal cycles [7], which makes the microstructure of weldment more complex, further impacting the performance of welding joints [8]. Fink [9] used strain-to-fracture (STF) tests to investigate relationships between cracking morphology and metallurgical factors for Ductility-dip cracking (DDC) in the heat-affected zone(HAZ) of NiCr15Fe-type alloys. On other hand, it is well known that welding is a complex process which influenced by many parameters [10], especially, heat input is a key factor determining the performance and morphology of the weldment. Chen et al. [11] studied the effects of texture and grain boundary (GB) misorientation on liquation cracking susceptibility. They found that both liquation cracking and crystal misorientation were increased as increasing the heat input. Ma et al. [12] revealed that the carbide precipitation is a major microstructural characteristic during heat treatment of stainless steels and nickel-based alloys. And Lee et al. [13] identified the GB carbides in Inconel 690 tubes as chromium rich (Cr-rich) $M_{23}C_6$ carbide by employing selected area electron diffraction (SAED) pattern analysis. Meanwhile, Blaizot et al. [14] used EDX chemical measurement revealed that their composition is $Cr_{23}C_6$. Another important thing is that Trillo and Murr [15] assessed the sensitization and precipitation behaviors of $M_{23}C_6$ on GB by using the transmission electron microscopy (TEM). They found that a few small precipitates nucleated on small angle GBs in 304 stainless steels, while larger amount of carbides were observed on high angle GBs. Hence, they concluded that $M_{23}C_6$ precipitates were prone to nucleate at the high angle GBs because of the highest energy. Therefore, according to the above mentioned researches, the heat input may affect the $M_{23}C_6$ carbides precipitation, which deserve further research.

The elements have very important influence on the precipitation. For instance, it has been determined that Al and Ta alloying elements enrich interdendritic regions, while W, Cr, Mo, and Co are concentrated in the dendrite branches reported by Ospennikova et al. [16]. Wang et al. [17] reported that Cr-C co-segregation at grain boundaries induced the precipitation of chromium carbides $M_{23}C_6$. Moreover, as the aging time was fixed, the GB precipitates $M_{23}C_6$ increased with the increase of GB angle. Since precipitations at GBs are important factor affecting the mechanical properties of metals [4], it is necessary to investigate the morphology, evolution process of precipitations at the GBs, as well as their effects on performance of metals. Especially, for Ni based alloys, which were usually served in high temperature, high pressure and corrosive atmosphere, their high temperature performance becomes more important. Moreover, due to alloying, precipitations at the GBs were the typical feature of the Ni based alloys, of course they had effect on performance of Ni based alloys. On other hand, during welding, the temperature field was not uniform, coupled with reheating cycle due to multi-pass welding, the microstructure, including the precipitations at the GBs, became more complex. Therefore, the evolution process of precipitations at the GBs of Ni based alloys with reheating thermal cycle needs to be well investigated, as well as its effect on mechanical properties.

In this study, hardfacing layers of Ni690 alloy were fabricated by gas tungsten arc welding (GTAW), with two different heat inputs, namely, large heat input (LHI) and small heat input (SHI). It is found that the microstructures of the two layers were different, especially, the proportion of high angle GBs (>15°) in the LHI sample was obviously higher than that in the SHI sample. And more $M_{23}C_6$ particles precipitated at the high angle GBs. The high temperature performance of Ni690 was investigated by employing Gleeble-3500 thermal/force simulation test machine. Microstructure was analyzed by optical microscope (OM), scanning electron microscope (SEM), and electron backscattered diffraction (EBSD). In particular, the influence of grain boundary precipitates $M_{23}C_6$ was investigated. The work provides the experimental and theoretical basis for the reasonable welding process of Ni690 alloy.

2 Materials and Experimental Procedure

The nominal chemical compositions in mass percent (wt%) of the deposited metal studied in this work are listed in Table 1. The metal was deposited on the base metal SA-508Gr.3Cl.1 by gas tungsten arc welding (GTAW). The welding parameters are shown in Table 2. The specimens for mechanical analysis were cut to tensile samples with thickness of 3 mm (Fig. 1), and the Gleeble-3500 thermal/mechanical simulation tester was employed to assess the mechanical properties at high temperature. Prior to thermal/mechanical tests, both sides of the specimens were pre-grinded till 400# abrasive paper. Then one side of the specimens was continually grinded till 2000# abrasive paper and polished with 5, 3, 1 μm diamond pastes in turn. After that, the sample thickness was reduced to about 2.8 mm. The process of high temperature tensile test was depicted in Fig. 2. After stretching at high temperature, the samples were chemically etched in 10% chromic acid solution for about 30 s with voltage of 12 V. Further, the surfaces were scrubbed with 2% dilute nitric acid corrosion, to remove the light-yellow corrosion layer. Then, the

Table 1 Chemical composition (mass percent) of deposited metal

Composition	C	Si	Mn	P	S	Cr	Ni
Content	0.03	0.26	3.99	0.005	0.0006	28.91	55.68
Composition	Mo	V	W	N	Al	Ti	B
Content	0.18	0.03	0.05	0.03	0.16	0.20	0.001
Composition	Co	Zr	Cu	Fe	Bi	Nb + Ti	Others (Co, B, Zr)
Content	0.03	0.003	0.03	8.79	0.001	1.60	0.08

Table 2 Welding parameters

No.	Welding parameters	Peak current (A)	Base current (A)	Duty cycle (%)	Welding speed (mm/min)	Wire feed speed (mm/min)	Welding voltage (V)
1	Large heat input (LHI)	250	180	40	140	1200	15–18
2	Small heat input (SHI)	300–320	240	40	150	1800	15–18

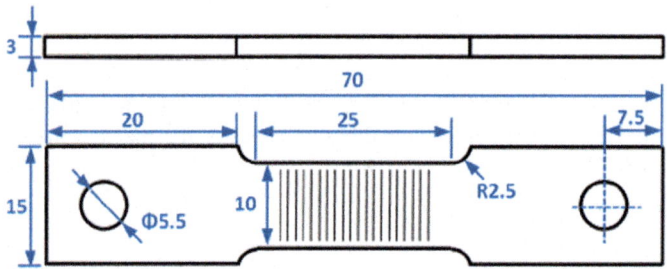

Fig. 1 Sketch map of high temperature tensile specimen

microstructures of the samples were investigated by using optical microscopy (AxioCam MRc5, Carl Zeiss) and scanning electron microscopy (NOVA NanoSEM 230, FEI). In order to analyze the GB angle distribution, the polished specimens were electro-polished in a solution of 10% $HClO_4$ + 90% CH_3COOH at room temperature with DC voltage of 15 V for 30 s, and characterized by electron backscattered diffraction (EBSD).

After reheating thermal cycle in Gleeble-3500 thermal/mechanical simulation tester, the tensile fracture surfaces were analyzed by using scanning electron microscope (SEM). And the microstructures of the specimen and morphologies of the $M_{23}C_6$ carbides were characterized in detail by OM and SEM.

Fig. 2 Schematic diagram of high temperature tensile test

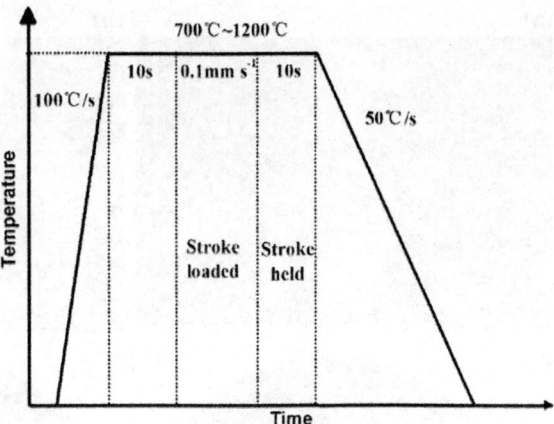

3 Results and Discussion

3.1 Influence of Heat Input in the Overlay Deposition

The supplied volume energy density E_v (in J/mm^3) can be used to evaluate the level of heat input for the cladding process, which plays a key role on the precipitations at GBs. The E_v that used by Yang et al. [18] is modified to measure average applied arc energy per unit volume during deposition:

$$E_V = \frac{U \times I}{S \times T \times V} \tag{1}$$

where U is the welding voltage (in V), I is the welding current (in A), S is the average pass space (in mm), T is the average thickness (in mm) of layer, and V (in mm/s) is the welding velocity. The area of layer was statistic by image J software, the calculated volume energy density E_v for large heat input (LHI) sample is about 559.65 J/mm^3. While for small heat input (SHI) sample, E_v is about 482.06 J/mm^3. Thus, the LHI deposition has a higher heat input per unit volume, relative to the SHI deposition.

3.2 Microstructure Characteristic of the as Surfaced Layers

Since the precipitations at GBs are affected by the high angle GBs, it is better to design samples with different high angle GBs. In this work, we found that the amount of high angle GBs can be changed by different heat input. Figure 3 depicts the results of EBSD characterization. EBSD inverse pole figure (IPF) maps observed from z direction show the grain crystallographic orientation and grain size

(a)   **(b)**

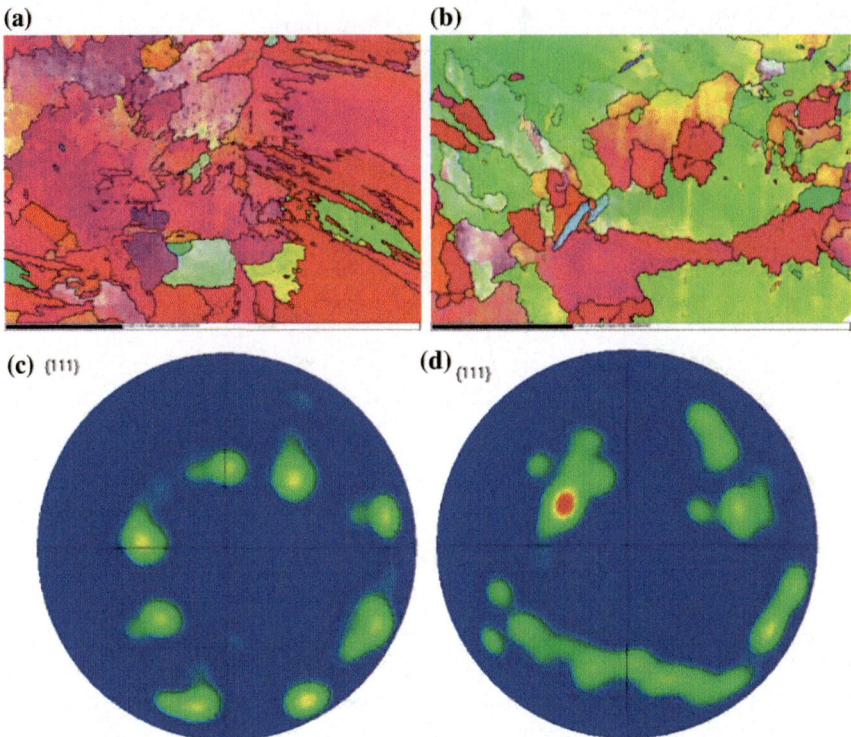

(c) {111}   **(d)** {111}

Fig. 3 Inverse pole figures (IPF) of **a** LHI and **b** SHI, corresponding pole figures (PF) of **c** LHI and **d** SHI

distribution. The pole figure (PF) in Fig. 3c, d indicates that orientation of the microstructure deviated from the <100> direction. And the grain orientation was more dispersed for the LHI sample.

The grain boundary angle distribution can be obtained from Fig. 4. It is found that there was obvious difference in the GB angle distribution. We divided the GB angle into three categories: <5°, 5–15° and >15°, which <5° was called as the small GB angle, the 5–15° was called as the middle GB angle, and that >15° was called as the high GB angle. There was no significant difference in the small GB angle and the medium GB angle. However, the proportion of high GB angle in the LHI samples was more than that in the SHI samples, as shown in Fig. 4. It is also proved that the heat input had a significant influence on the GB angle distribution [11].

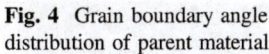

Fig. 4 Grain boundary angle distribution of parent material

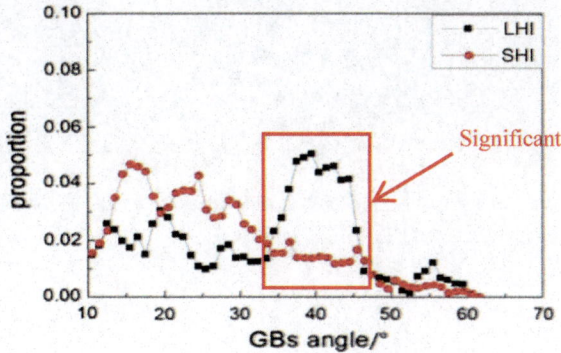

3.3 Effect of GB Carbides $M_{23}C_6$ on High Temperature Performance

To investigate the effect of carbides on the high temperature mechanical property, the evolution process of precipitations at GBs was analyzed by SEM, the results were shown in Fig. 5. By using selected area electron diffraction (SAED) pattern analysis, Lippold and Nissley [7] detected chromium rich (Cr-rich) $M_{23}C_6$ carbides on the GBs of Inconel 690 Tubes. Moreover, the crystallography relationship between the $M_{23}C_6$ carbide and matrix was expressed as $<100>M_{23}C_6//<100>$ matrix, $\{100\}M_{23}C_6//\{100\}$ matrix [19]. It is seen that, according to Fig. 5a–c, the morphology of $M_{23}C_6$ in the LHI sample changed a lot after a high-temperature cycle. Continuous $M_{23}C_6$ distributed at the GBs as the reheating temperature was 700 °C, which almost fully covered the GBs, as shown in Fig. 5a. As the temperature was increased to 900 °C, $M_{23}C_6$ particles started to dissolve, and distributed semi-continuously, as shown in Fig. 5b. While as the temperature was 1050 °C, most $M_{23}C_6$ particles disappeared from the GBs, as shown in Fig. 5c. The results indicate that the amount of precipitations at GBs decreased with increase of reheating temperature. Figure 5d–f shows the $M_{23}C_6$ morphology in the SHI sample after a high-temperature cycle. The evolution process of precipitations at the GBs of the samples produced with SHI exhibited the similar rule, that is, the precipitations at GBs gradually dissolved with temperature increase at range of 700–1050 °C.

Meanwhile, according to the SEM photograph of GB, the amount of precipitates at the GBs of samples produced with LHI was much higher than those samples produced with SHI. For quantitative comparing the proportion of precipitates at GBs, image J software was employed. It can be seen from the Fig. 6 that the proportion of the GB precipitates in the LHI samples was 75%, and that in the SHI samples was about 49%, it is obviously lower than that of LHI samples. Moreover, this result is consistent with the comparison in the amount of the high angle GBs in the both samples produced with LHI and SHI, respectively. In other word, the GB $M_{23}C_6$ precipitates were more likely to precipitate at the high angle GBs. Lee et al.

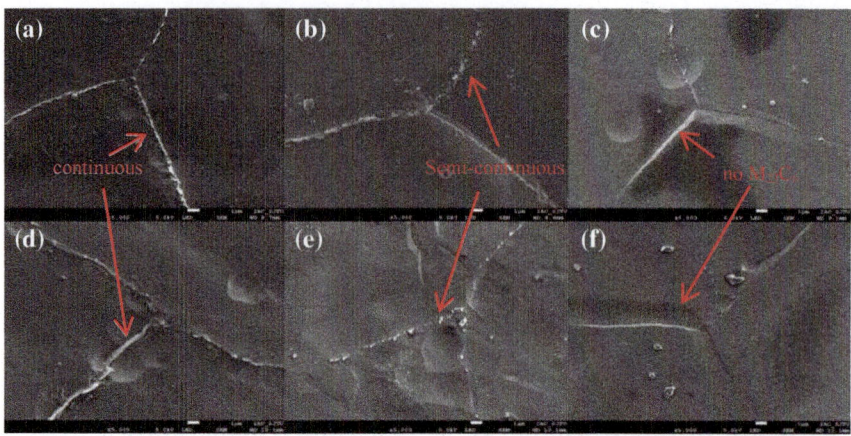

Fig. 5 Size and distribution of precipitates **a** LHI 700 °C; **b** LHI 900 °C; **c** LHI 1050 °C; **d** SHI 700 °C; **e** SHI 900 °C; **f** SHI 1050 °C

[13] also found that as the aging time was fixed, the GB precipitates $M_{23}C_6$ increased with the increase of GB angle. In this work, $M_{23}C_6$ was distributed in skeleton form at 700 °C, because the LHI samples had a large proportion of high angle GBs. The enhancement of the strength depends on the interaction of GBs, grain and GB $M_{23}C_6$ precipitates [10]. Moreover, $M_{23}C_6$ can effectively pin the GBs against migration, further improving the high temperature strength.

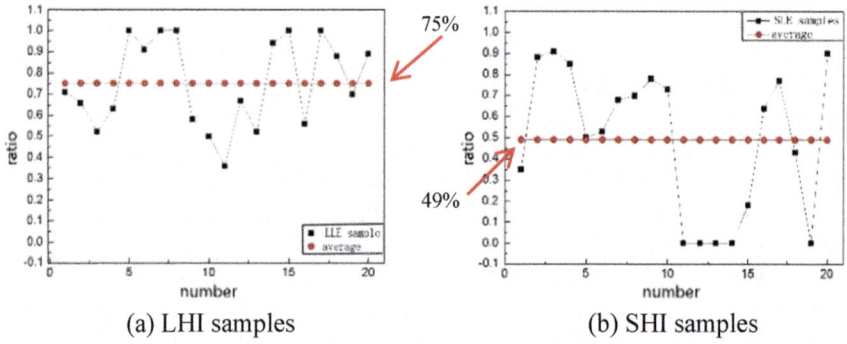

(a) LHI samples   (b) SHI samples

Fig. 6 Grain boundary precipitation ratio of the as surfaced samples

3.4 Effects of Difference Heat Input on High Temperature Performance

Figure 7 plots the stress-strain curves of the samples during reheated to temperature of 700–1200 °C. It is found that, after yielding, the shape of stress-strain curve of the samples at 700 and 800 °C was very different from that at 900–1200 °C, regardless the heat input. The stress-strain curves of samples reheated to 700 and 800 °C exhibited a remarkable strain hardening feature. And they displayed a longer extension and abrupt fracture at last. However, the stress-strain curves of the samples after reheated to 900–1200 °C, leveled off after yielding, and elongation percentage was remarkably reduced, especially for the samples manufactured with SHI. There was a gradual fall in stress at the end of the tensile. The reason is that when the temperature was below 900 °C, the grain boundary strength was higher than that in the crystal, and the deformation was mainly concentrated in the crystal; however, when the temperature was higher than 900 °C, the grain boundary strength was weaker than the intragranular strength. At the same time, grain boundary sliding and dislocation climb became the main deformation mechanism [14].

Figure 8 presents the ultimate tension strength (UTS) as a function of reheating temperature. It is obvious that the high temperature UTS decreased with the increase of the reheating temperature. Moreover, the UTS of samples produced with LHI was larger than that with SHI at the corresponding reheating temperature. As can be seen from Fig. 8, the maximum UTS difference between two layers occurred at 700 °C, which was 33.79 MPa. However, the difference became smaller with the increase of temperature. As temperature ranging from 1000 to 1200 °C, the difference of the ultimate tensile strength value was not large.

Combining with the microstructure analysis, both the change of UTS with temperature, and the difference of UTS between different samples, had consistent trend with the evolution process of precipitates at the GBs. The amount of $M_{23}C_6$ at the GBs of LHI samples was higher than that of SHI samples, improving the effect of grain boundary pinning. Therefore, the ultimate tensile strength of LHI samples

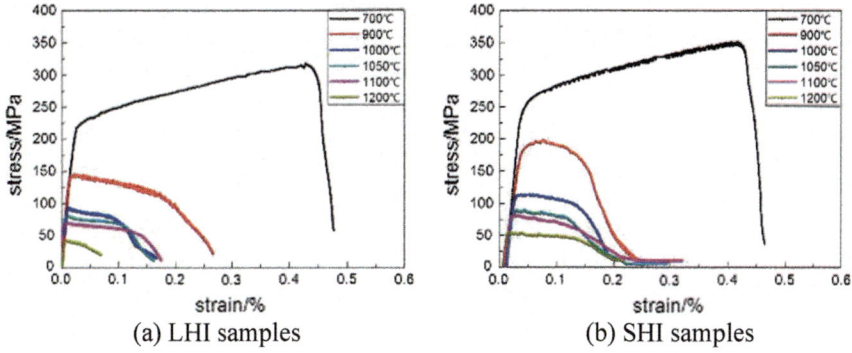

(a) LHI samples   (b) SHI samples

Fig. 7 High temperature tensile stress-strain curves of LHI and SHI samples

Fig. 8 Comparison of the ultimate tensile strength of LHI and SHI samples at various temperatures

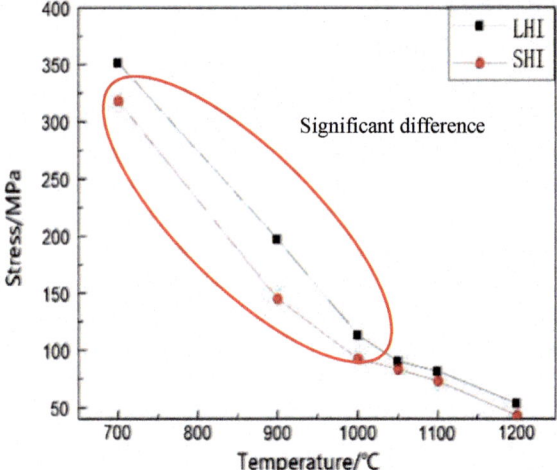

was higher than that of the SHI samples. However, when the temperature was at the range of 1000–1200 °C, $M_{23}C_6$ had been dissolved into the matrix, which lead to the pinning effect on grain boundary be weakened, so it had little effect on the high-temperature ultimate tensile strength.

3.5 Fracture Morphology Analysis

The transverse section of the fracture was observed by the optical microscope, the samples which experienced reheating to 700, 900, and 1050 °C were selected to analyze the fracture characteristic. It is found that the fracture modes of the LHI samples were the same as the SHI samples, as shown in Fig. 9. As the reheating temperature was 700 °C, the failure occurred in a transgranular mode; as the reheating temperature increased from 900 to 1050 °C, the fractures for the high temperature tensile test turned from intergranular mode to transgranular mode. Previous research [20] had shown that when the temperature was between T_{E1} to T_{E2} (T_{E1}: 650–810 °C; T_{E2}: 870–930 °C), the GB strength was weaker than the intragranular strength, at this time, the fracture was transgranular brittle fracture. On the other hand, when the temperature was below T_{E1} or higher than T_{E2}, the grain boundary strength was higher than that of intragranular strength, which was intergranular ductile fracture.

Scanning electron microscopy (SEM) was used to observe the fracture morphology (700, 900, 1050 °C), and the fracture morphologies were shown in Fig. 10. It can be seen from Fig. 10a that the fracture surface exhibited a fine dimpled surface at 700 °C, which is the typical characteristic of the dimpled ductile mode of failure. It is attributed to the decrease of chromium concentration in the matrix which due to the coarse GB $M_{23}C_6$ carbides precipitated at the grain

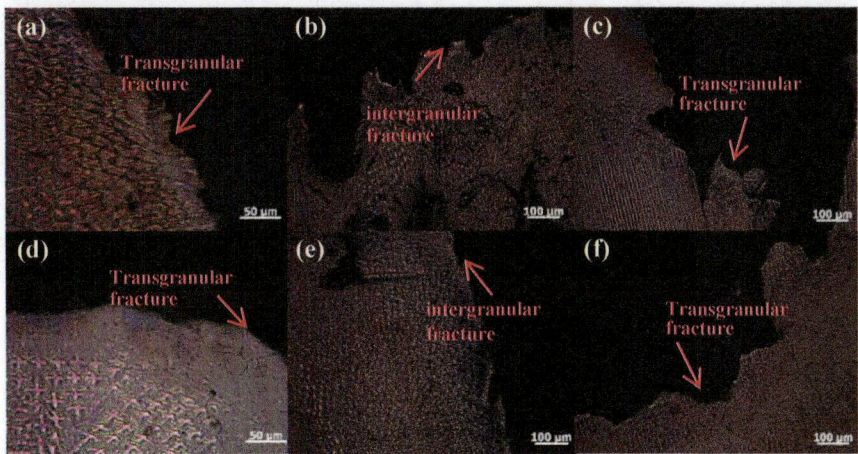

Fig. 9 Metallograph of the transverse section of the fracture **a** LHI 700 °C; **b** LHI 900 °C; **c** LHI 1050 °C; **d** SHI 700 °C; **e** SHI 900 °C; **f** SHI 1050 °C

boundary. It indicates that the strength of the GB was higher than that of the crystal due to the dislocation slipped to GBs; as temperature was 900 °C, it was brittle intergranular fracture, while dimple fracture disappeared. Meanwhile, the maximum strength difference between grain and GB was generated at 900 °C (Fig. 8), for which the fractures were entirely intergranular fractures (Fig. 10b, e). At this temperature, a higher content of $M_{23}C_6$ particles precipitated at the GBs (Fig. 5b, e). The intergranular fracture was caused by grain boundary embrittlement. As the temperature was 1050 °C, it was transgranular fracture and had many shallow dimples. Corresponding to the metallographic photograph (Fig. 9), the fractures of the high-temperature tensile test turned from dimple fractures to intergranular fractures at temperatures ranging from 700 to 900 °C, and then turned to dimple fracture again as the temperature continuously increased to 1050 °C. This result shows that the strength of the grain boundaries and crystal changed with the increase of temperature. At the beginning, the strength of the crystal was weaker than the GBs because of the precipitations, so the crack was easy to grow and expand in the crystal. Previous research had shown that when GB carbides in super alloys were continuously presented along the GBs, it would be deleterious to intergranular crack growth resistance [21]. With the increase of temperature, the strength of GBs decreased faster than that crystal due to dissolution and precipitation of precipitations, resulting in initiation of crack at GBs. As the temperature continued to rise, the intergranular carbides dissolved completely, leading the fractures turn to dimples, and the GBs strength is higher than that of crystal. Lee et al. [13] investigated the influence of GB precipitates behavior of Inconel 690 tube on tensile property, they found that $M_{23}C_6$ easily precipitated in the high angle GBs because of high surface energy. With the increase of aging time, the tensile strength and elongation decreased significantly. By analyzing the fracture surface, it is found that the precipitate $M_{23}C_6$ was the cause of crack initiation [20].

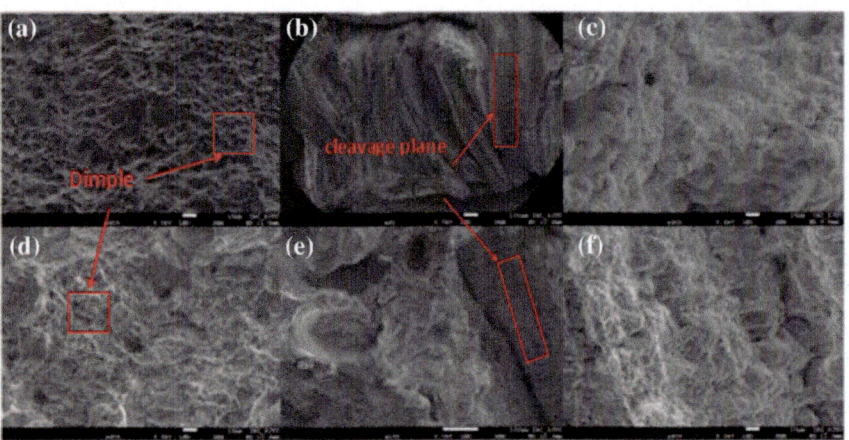

Fig. 10 Fracture morphology **a** LHI 700 °C; **b** LHI 900 °C; **c** LHI 1050 °C; **d** SHI 700 °C; **e** SHI 900 °C; **f** SHI 1050 °C

4 Conclusion

In present work, large heat input (LHI) and small heat input (SHI) were separately used to fabricate Ni 690 alloy hardfacing layer. The effect of GB angle and GB precipitates $M_{23}C_6$ on high-temperature performance were investigated. The major conclusions are listed as follows:

1. By changing the heat input, two kinds of hardfacing layer with obvious difference in grain boundary angle distribution were obtained. The samples produced with LHI had more large angle GBs than that produced with SHI.
2. The ultimate tensile strength of LHI samples was higher than that of SHI samples at the temperature range of 700–1000 °C. Moreover, it had significant difference at 700 and 900 °C. While as the temperature increased, the difference became smaller. At the range of 1000–1200 °C, the values of UTS were slightly different.
3. At 700 °C, continuous $M_{23}C_6$ particles distributed at the GBs, which almost fully covered the GBs; when the temperature increased to 900 °C, $M_{23}C_6$ became larger and dispersed at the GBs; while as the temperature was over 1050 °C, $M_{23}C_6$ particles dissolved. The evolution process of precipitations at the GBs of the samples produced with SHI and LHI exhibited the similar rule.
4. In SHI samples, the proportion of the large angle GBs was much higher than that of SHI samples, $M_{23}C_6$ was easier to precipitate at the large angle GBs. And the $M_{23}C_6$ at the GBs can effectively pin the grain boundaries against the migration, and improve the high temperature strength.
5. The fracture modes of the both samples were same: at 700 °C, it was transgranular fracture; at 900 °C, it was intergranular fracture; and at 1050 °C, it turned to transgranular fracture.

Acknowledgements The authors gratefully acknowledge the financial funding of the National Natural Science Foundation of China (51204107, 51575347 and 51405297).

References

1. Qin R, Wang H, He G (2014) Investigation on the microstructure and ductility-dip cracking susceptibility of the butt weld welded with ENiCrFe-7 nickel-base alloy-covered electrodes. Metall Mater Trans A 46(3):1227–1236
2. Lu Z, Chen J, Shoji T et al (2015) Characterization of microstructure, local deformation and microchemistry in Alloy 690 heat-affected zone and stress corrosion cracking in high temperature water. J Nucl Mater 465:471–481
3. Santra S, Rao Ramana, Kapoor K (2016) The effect of thermal treatment on microstructure and corrosion of nickel base Alloy 690 (UNS N0 6690). Mater Perform Charact 5(1):20160046
4. Chen JQ, Lu H, Cui W et al (2014) Effect of grain boundary behavior on ductility dip cracking mechanism. Mater Sci Technol 30(10):1189–1196
5. Lin CM (2013) Relationships between microstructures and properties of buffer layer with Inconel 52M clad on AISI 316L stainless steel by GTAW processing. Surf Coat Technol 228:234–241
6. Lee HT, Chen CT (2011) Predicting effect of temperature field on sensitization of alloy 690 weldments. Mater Trans 52(9):1824–1831
7. Lippold JC, Nissley NE (2008) Ductility-dip cracking susceptibility of nickel-based weld metals Part 1: strain-to-fracture testing. Weld J 87:257–264
8. Shen RR, Strom V, Efsing P (2016) Spatial correlation between local misorientations and nanoindentation hardness in nickel-base alloy 690. Mater Sci Eng, A 674:171–177
9. Fink C (2016) An investigation on ductility-dip cracking in the base metal heat-affected zone of wrought nickel base alloys—part I: metallurgical effects and cracking mechanism. Weld World 60(5):939–950
10. Zheng L, Hu X, Kang X et al (2015) Precipitation of $M_{23}C_6$ and its effect on tensile properties of 0.3C-20Cr-11Mn-1Mo-0.35N steel. Mater Des 78:42–50
11. Chen Y, Zhang K, Huang J et al (2016) Characterization of heat affected zone liquation cracking in laser additive manufacturing of Inconel 718. Mater Des 90:586–594
12. Ma Y, Li S, Hao X et al (2016) Research on the carbide precipitation and chromium depletion in the grain boundary of alloy 690 containing different contents of nitrogen. Acta Metall Sin 52(8):980–986
13. Lee T-H, Lee Y-J, Joo S-H et al (2015) Intergranular $M_{23}C_6$ carbide precipitation behavior and its effect on mechanical properties of Inconel 690 tubes. Metall Mater Trans A 46(9):4020–4026
14. Blaizot J, Chaise T, Nelias D et al (2016) Constitutive model for nickel alloy 690 (Inconel 690) at various strain rates and temperatures. Int J Plast 80:139–153
15. Trillo EA, Murr LE (1998) A TEM investigation of $M_{23}C_6$ carbide precipitation behaviour on varying grain boundary misorientations in 304 stainless steels. J Mater Sci 33(5):1263–1271
16. Ospennikova OG, Petrushin NV, Treninkov IA et al (2016) Phase and structural transformations in heat resistant intermetallide nickel-based alloy. Inorg Mater: Appl Res 7(6):832–839
17. Wang M, Zha X, Gao M et al (2015) Structure, microsegregation and precipitates of an alloy 690 ESR ingot in industrial scale. Metall Mater Trans A 46(11):5217–5231
18. Yang J, Li F, Wang Z et al (2015) Cracking behavior and control of Rene 104 superalloy produced by direct laser fabrication. J Mater Process Technol 225:229–239

19. Lim YS, Kim DJ, Hwan SS et al (2014) $M_{23}C_6$ precipitation behavior and grain boundary serration in Ni-based Alloy 690. Mater Charact 96:28–39
20. Mo W, Lu S, Li D et al (2014) Effects of $M_{23}C_6$ on the high-temperature performance of Ni-based welding material NiCrFe-7. Metall Mater Trans A 45(11):5114–5126
21. Defects W (2015) Effect of Nb content of Nb content on microstructure, welding defects and mechanical properties of NiCrFe-7 weld metal. Acta Metall Sin—Chinese Edition 51(2): 230–238

Migration Behavior of Tungsten Carbide in the Dissimilar Joints of WC-TiC-Ni/304 Stainless Steel Using Robotic MIG Welding

Guotao Ying, Hongying Gong and Peiquan Xu

Abstract Using robotic metal inert gas (MIG) welding, WC-TiC-Ni/304 stainless steel was fabricated using pure nickel as welding wire. The welds consisted of the austenitic γ-Ni matrix, dissolved WC, and compound carbide (W, M)C. Electromagnetic stirring-induced (type I), diffusion-induced (type II), and shear-induced (type III) WC migration led to WC_a and WC_b type migration. The gradient layer and η-phase were formed at the interface. WC_a migration (type I and type II) showed that arc plasma provided enough energy for WC_a long-range migration from the heat affected zone (HAZ) to the fusion zone. WC_b migration (III) exhibited the stress levels to be above the yield stress in the fusion zone. A self-sealing model was put forward to describe WC migration and the formation of gradient layers. The results showed that WC migration not only occurred in the fusion zone, but also in the HAZ, especially near the top surface, which led to gradient layer, η phase and tungsten dissolution-re-precipitation on the surface of WC. The results also indicated that the fusion zone had the ability to cure the cracks itself during the robotic MIG welding.

Keywords Robotic MIG welding · Electromagnet-induced migration
Diffusion-induced migration · Shear-induced migration · Gradient layer

1 Introduction

Among ceramics, carbides of transition metals such as WC, TiC or TaC are refractory compounds and possess high hardness and strength, which are maintained even at elevated temperatures [1]. Especially WC-based cermet has been widely used in military, aerospace, automotive, marine, petrochemical, mining, electronics, microelectronics and wood industries [2] since decades in various

G. Ying · H. Gong · P. Xu (✉)
College of Materials Engineering, Shanghai University of Engineering Science,
Shanghai 201620, China
e-mail: pqxu@sues.edu.cn

© Springer Nature Singapore Pte Ltd. 2018
S. Chen et al. (eds.), *Transactions on Intelligent Welding Manufacturing*,
Transactions on Intelligent Welding Manufacturing,
https://doi.org/10.1007/978-981-10-8330-3_9

engineering applications, such as pipe-valve component, cutting tools, drill tips and as well as various wear-resistance parts [3–5]. However, WC dissolution phenomenon usually happened during liquid phase sintering of WC-Co alloys, fusion welding of WC-Co/ steel or WC laser clad. During the dissolution, the alloys, welds or clad always display a skeleton of WC grains embedded in a Co/Fe rich matrix [6] and then the structures loss the original densification. In addition, WC dissolution varies with temperature and forces: brim dissolution; transcrystalline rupture; aggregate coarsening; large region yield; carbon or tungsten diffusion from inner of WC to WC/β interface; iron diffusion from matrix to WC [7–10], which deteriorates the original microstructure and mechanical properties. Welding is accepted for joining cemented carbide for simple applications, and likewise for those involving the most complicated critical structures where failure could be destructive to life and property. The welded joint is designed to meet a certain combination of properties required by the end use. Thus, the resulting mechanistic understanding of WC dissolution will be used to improve the weldability of cemented carbide during dissimilar welding of cemented carbide and steel and provide a scientific basis for cemented carbide-steel application in "next generation" oil and gas transportation equipment under conditions of technological importance.

Most of the important, nano-WC based cermet is under development to solve the problem of the low lifetime of cermet product because of the beneficial effect of the size and the shape of WC crystals on the mechanical properties [2, 11–13]. Nano-WC is the most reactive one, higher specific surface areas enhance welding activity and recrystallization process of WC grains, therefore, the extensive growth and dissolution of WC nano-grains easily occur during welding.

Besides WC dissolution in liquid Co/Fe matrix, element diffusion also leads to WC dissolution in solid phases accompanied by liquid dissolution [14–19]. In generally, mechanical alloying is a solid state processing technique involving repeated welding, fracturing, and rewelding of powder in high energy ball mill [20, 21], which is a good choice to investigate WC dissolution in the solid W-Fe-C system. Many studies [15, 18, 22, 23] also revealed that the element additives had a strong effect on the WC dissolution in solid phases during element diffusion. These additives included carbon, nickel, iron, rare earth, chromium, and something else. The main role of RE additives was addressed during sintering when their presence at the boundary of WC induced oxide dispersion strengthened effectiveness and inhibition of recrystallization process of WC grains [11, 24]. It was also concluded that an effective control of the reaction can be achieved by increasing in the binder composition the ratio between the elements without affinity and those with affinity to carbon and/or enlarging the carbon content [25]. Thus, the size of WC crystals had a strong effect on the mechanical properties of WC/Co. Research results [13, 14] also indicated that mechanical properties were influenced by the shape of WC crystals.

The present research aims to investigate WC migration during robotic MIG welding. Consequently, WC migration through transition layer based on thermo-dynamics and WC migration in the liquid W-M-C system are discussed.

2 Experimental

WC-TiC-Ni used in present study has a chemical composition (wt%) of 15 TiC, 79 WC, and balance Ni. Hard phases WC, TiC are embedded in the binder nickel matrix. The chemical composition (wt%) of 304 stainless steel is 0.08 carbon, 2.0 manganese, 0.045 phosphorus, 0.03 sulfur, 1.0 silicon, 18.2 chromium, 10.1 nickel, and balance iron. Nickel wire is used as filler metal and has a chemical composition (wt%) of 0.05 carbon, 0.1 manganese, 0.15 silicon, 0.1 iron, 0.1 cobalt, and 99.5 nickel, impurity does not exceed 0.5, and the diameter is 1.2 mm. The base materials were cut into plate with a square section of 200 mm × 100 mm and 4-mm thick with the 30°, 45° and 60° grooves respectively.

As shown in Fig. 1, welded joints were prepared using GLC 603 QUINTO welder with single wire, welding positioner, CLOOS@350 robot manipulator with control ROTROL II. With argon, the metal transfer mode, for the range of welding conditions investigated in present research, is normally globular transfer and there is no transition to spray transfer. The parameters for robotic MIG welding were shown in Table 1.

Heat input is a relative measure of the energy transferred per unit length of weld. It is typically calculated as the ratio of the power to the speed of the arc as follows,

$$H = (60 \cdot Q)/(1000 \cdot S) \tag{1}$$

where, H is the heat input (kJ/mm), S is the travel speed (mm/min), and the rate of heat transfer Q from the arc to the workpiece is determined by the following equation [26],

$$Q = \eta \cdot Q_{nominal} \tag{2}$$

where η is arc efficiency, here, η is equal to 0.8, $Q_{nominal}$ is the nominal power of the arc, $Q_{nominal} = E \cdot I$, where E is arc voltage, I is welding current. With GLC 603 QUINTO welder, E values are set before welding, and pick I values from the indicator.

Fig. 1 Architecture of robotic MIG welding of cemented carbide to 304 stainless steel

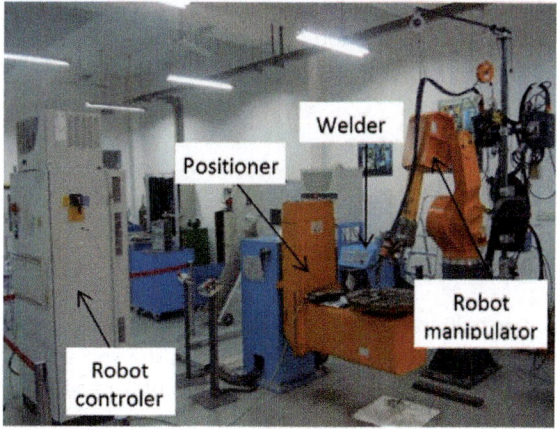

Table 1 Robotic MIG welding parameters and weld defects in present study

d (mm)	No.	Heat input (J/mm)	Groove (°)	Max. d_1 (mm)	Weld defects	Crack
4	A1	321	30	42.98	Cracks	30 s after welding
	A2	406	30	56.94	Cracks	5 s after welding
	A3	463	30	136.94	Cracks	2 min after welding
	A4	417	30	103.39	–	–
	A5	411	30	95.77	–	–
6	I-1	418 (416)	30	143.05	Spatter, porosity	No cracks (2 s)
	I-2	429	30		Insufficient	10 s after welding
	I-3	392	30		Insufficient	5 s after welding
	II-1	432 (367)	45	120.8	Insufficient	No cracks (instant)
	II-2	439 (459)	45		Insufficient	No cracks (instant)
	II-3	420 (363,377)	45		–	No cracks (instant)
	III-1	529	60	186.68	Burn-through	During weld
	III-2	566 (517)	60		EP	During weld
	III-3	441	60		Insufficient	6 h after welding

Tungsten carbide is hexagonal with lattice constants a = 2.91 Å and c = 2.84 Å. If sufficient C and W dissolved into the Co, the phase is stabilized [27]. The microstructure was characterized by optical metallography (OM), electro-probe micro analyzer (EPMA) and scanning electron microscopy (SEM). SEM images were performed using S-4800 (Hitachi, Japan). The specimens were prepared by the standard metallographic procedures and etched with the Murakami's reagent [28]. The reagent ferric chloride (3 g $FeCl_3$ and 100 ml H_2O) was used for nickel binder removal (10 s).

3 Results

3.1 Weld Formation

For metallographic examination, the specimens were cut vertical to welded seam. After grounded, polished and etched, the specimens were then investigated along the position of cross-section. Figure 2 illustrated the influence of thickness and heat input on the weld appearance of WC-TiC-Ni/304 joint. The result (Table 1) indicated the use of higher heat input (A3: 463 J/mm) led to wider fusion zone compared to lower heat input (A1: 321 J/mm; A2: 406 J/mm). Cracks were observed through the depth in specimen A1 and A2 at a minimum of 20, 5 s after welding. The cracks occurred at lower temperature, initiated from the HAZ near cemented carbide at the start and then propagated to cemented carbide. The average length of longitudinal cracks was 43 and 49 mm, and the average length of transverse cracks was 12 and 11 mm respectively. When the use of medium heat input (A4: 417 J/mm, A5: 411 J/mm), the consequent residual stress could not meet the requirement of crack initiation and propagation, cracks were not observed in specimens A4 and A5. However, with heat input H increased continuously (A3: 463 J/mm), the cracks occurred. Close to the top surface of welded joint, the welds were sensitive to transverse cracking. With the increase of heat input, more transverse cracks appear to occur on the surface of the bead. The cold cracks, when it happened, always occurred in the HAZ near cemented carbide; however, the hot cracks, when it happened, always occurred in the welds, on the top surface of previous bead, and then they propagated in transverse direction.

When 6-mm thick base materials were used, the cracks were always observed on the WC-TiC-Ni side. With the increase of groove angle, more and more filler metals were needed to fill the groove, the cracking susceptibility increased accordingly. When groove angle was equal to 30°, spatter, porosity or insufficient penetration were easily observed. Especially, two layers were adopted, no cracks occurred when P = 418 J/mm for first layer, however, cracks were propagated just after welding. When groove increased to 45°, two layers could not meet the requirement of weld penetration. When three layers were adopted, excess penetration (EP) made the weldment fail. EP, even burn through occurred when heat input increased to 500 or three layers were selected for groove angle 60°.

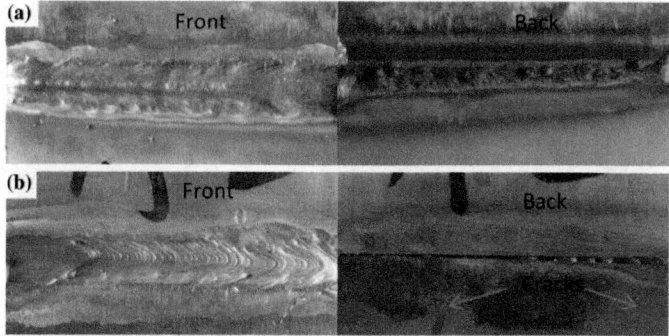

Fig. 2 Comparison of weld appearance between **a** d = 4 mm and **b** d = 6 mm (left: front; right: back)

3.2   Microstructures

For further investigation, the transverse cross-sections of 4 and 6-mm thick specimens were conducted in Figs. 3, 4 and 5. Cross-sectional view of typical joint A1
was illustrated in Fig. 3. WC migration and grain growth: normal grain growth
(NGG), abnormal grain growth (AGG) were observed near WC-TiC-Ni/welds
interface in both 4 and 6-mm thick specimens. In current research, an abnormal
grain was defined to be around seven times larger than the average grain size [29].
The WC migration from cemented carbide to weld, region 'A' in Fig. 3, was
amplified in Fig. 4a. The distance from original location to current position was
nearly 200 μm.

The WC migration, always accompanied by grain growth, occurred near the top
surface because of larger heat input and forces during welding. Especially, they
were affected by weld pool mixed with strong convection currents resulting from
the interaction between the electromagnetic and Marangoni forces. Far from 304/
fusion zone interface, equiaxed austenite grains and annealing twins were observed
(Fig. 5c), however, the weld was embrittled by $\alpha(\delta)$ ferrite $\rightarrow$ σ phase at 820 °C or
$\alpha(\delta)$ ferrite $\rightarrow$ α ferrite with low chromium (Fig. 5d) in HAZ at 304 stainless steel
side. Gradient microstructure in heat affected zone was observed not only in 4-mm
thickness specimens (Fig. 4c) but in 6-mm thickness specimens (Fig. 5a).
Coarse WC grains and fine dispersive WC grains were observed in both weld and
HAZ near WC-TiC-Ni/Ni interface. Coarse WC grains were divided into two parts
(Fig. 4d): NGG and AGG, which were also observed in 6-mm thick specimens
(Fig. 5e).

NGG was owed to the WC nucleation and grain growth in liquid nickel during
welding or final solidification stage; however, AGG was an abrupt growth phenomenon at elevated temperature and WC grains had a strong tendency for AGG
because of the difference of growth rate for different nucleation of growth ledges,
mass transfer across the interface and long range diffusion. Both grain growth

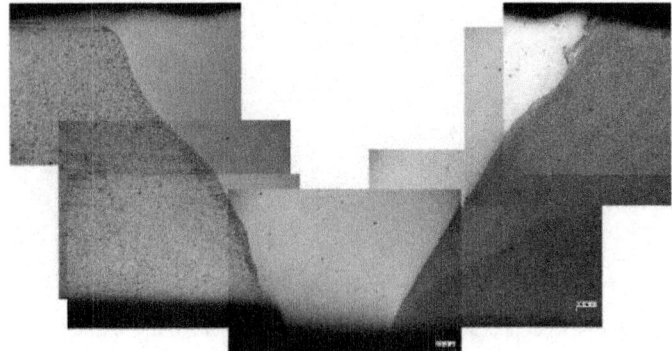

Fig. 3 Cross-sectional view of a typical joint (A1) of 304 stainless steel (left side) and
WC-TiC-Ni (right side) with pure nickel welding wire ($d = 4$ mm)

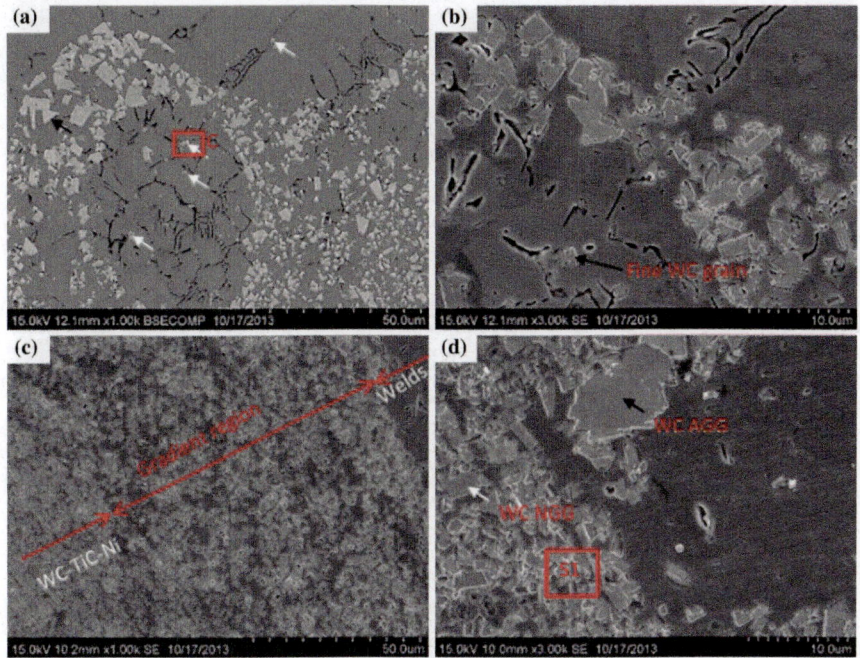

Fig. 4 WC behaviors near WC–TiC–Ni/weld interface in 4-mm thick specimens. **a** Migration of WC grains from cemented carbide to weld in specimen A1 (region 'A' in Fig. 3), white arrow showing fine WC grains; black arrow showing coarse WC grains; **b** WC phase surrounded by nickel from both binder and filler metal (region 'C' in **a**); **c** Gradient microstructure in specimen A5; **d** WC NGG and WC AGG observed in specimen A4

behaviors were strongly influenced by the initial size distribution [29, 30]. By contrast, 6-mm thick specimens had much stronger tendency for WC migration and grain growth than 4-mm thick specimens.

3.3 Morphology and Compositions

SEM images of the fusion zone and WC-TiC-Ni/welds interface were illustrated in Fig. 6, taken near the center of the cross section of the weldment. Figure 6a illustrated SEM images of WC-TiC-Ni/welds interface in specimen A1, white arrows showing grain boundaries, and black arrows showing the nickel grains. The morphology was characterized using backscattered electron imaging and secondary electron imaging respectively. The spectrums used for composition analysis were marked with yellow cross or yellow rectangular. And the corresponding composition and EDS spectrum were listed in Table 2. Dissolved WC embedded in nickel matrix, nickel mixtures of original nickel in WC-TiC-Ni and nickel from the filler

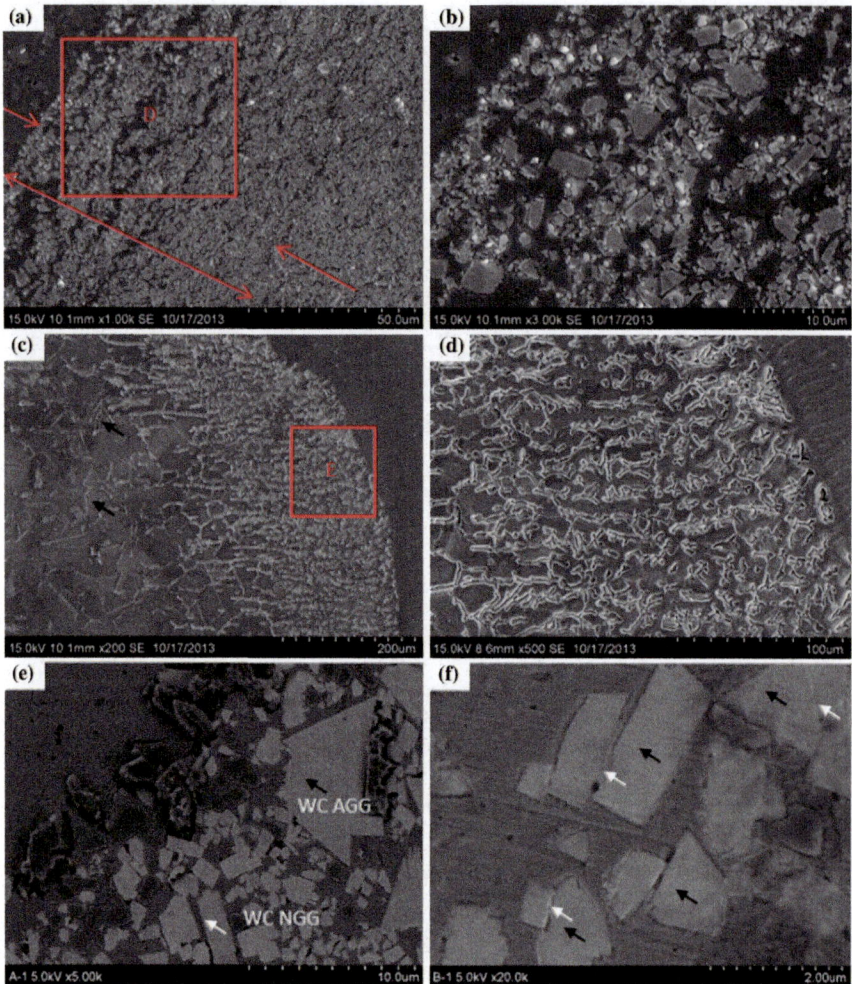

Fig. 5 SEM image showing the WC behaviors in 6-mm thick specimens. **a** Gradient microstructure in I-2 specimen; **b** Gradient WC-TiC-Ni accompanied by grain growth (region 'D' in **a**); **c** (Fe, Cr)/Ni solution near steel side in specimen I-2, black arrows indicate the original equiaxed austenite grains and annealing twins; **d** High magnification of region 'E' in **c**. **e** WC abnormal grain growth observed in specimen I-2; black arrow indicates AGG WC, white arrow indicates NGG WC. **f** WC migration to weld in specimen II-2; black arrows indicate the dissolved WC grains; white arrows show the interface WC/WC

metal. The grain size of nickel in welds was larger than that in HAZ near WC-TiC-Ni/welds interface. The dissolved WC indicated blunt corner. And the sharp triangle WC or rectangular WC grains were not observed. Results on composition analysis of WC-TiC-Ni/welds manifested that "pro-dissolved carbide" (SP1) had the average content (wt%) 16.22W, 43.61Ni, 21.52Fe, 1.22C, 9.45Ti and

5.36%Cr characterized with tungsten depletion. In contrast to weld, the WC and TiC were surrounded by nickel and iron liquid. The "pro-dissolved carbide" meant WC dissolution in nickel matrix was a gradual process until WC disappeared in the matrix completely.

However, it was difficult for pro-dissolved WC to transit from WC $\rightarrow$ η phase (Fe_3W_3C or Fe_6W_6C), the driving force for growth of "pro-dissolved carbide" grain was not enough, which made it impossible for nickel or iron atoms substitute for tungsten atoms. Therefore, WC $\rightarrow$ η phase transition was inhibited. Figure 6b

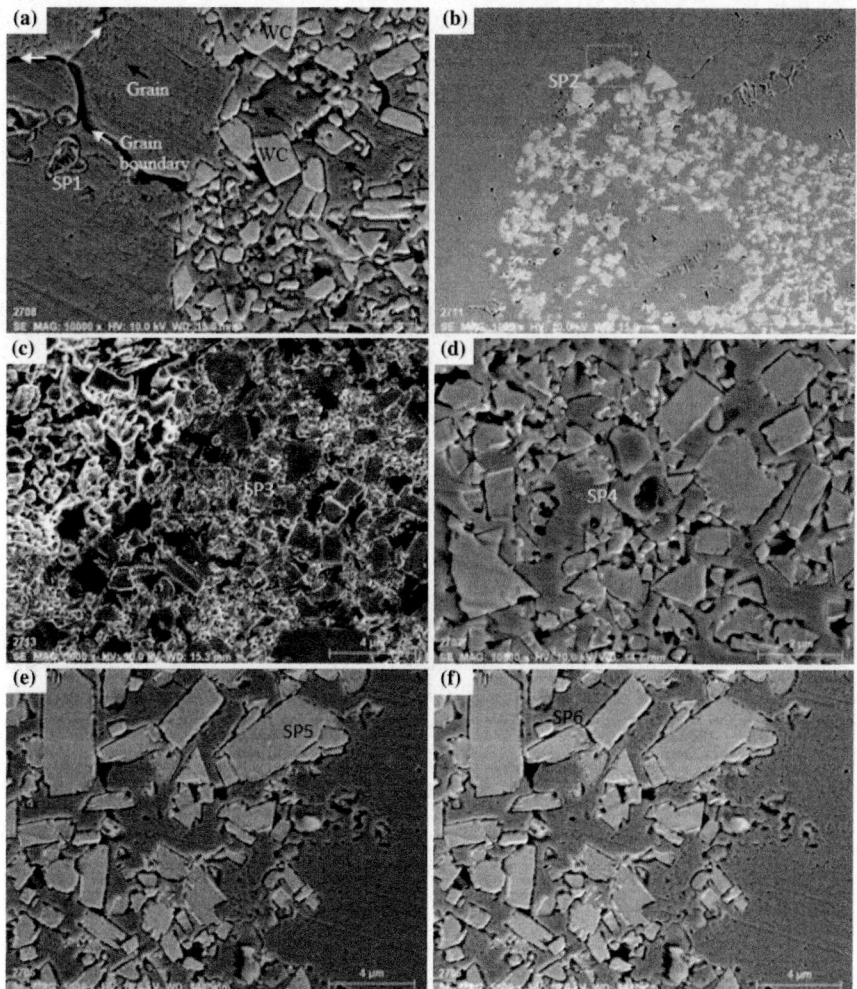

Fig. 6 SEM image showing the WC dissolution in 4-mm thick specimens, **a** SP1 in specimen A1, white arrow showing grain boundary; black arrows showing the nickel matrix; **b** SP2 in specimen A1; **c** SP3 in specimen A4; **d** SP4 in specimen A5; **e** SP5 in specimen A5; **f** SP6 in specimen A5

showed another WC dissolution in non-equilibrium η phase layer, was near the fusion line "non-equilibrium η phase" (SP2) had the average content (wt%) 40.50W, 44.71Ni, 12.91Fe, 1.87C characterized with carbon depletion. WC → η phase transition occurred and Fe₃W₃C compound was observed. This meant WC had experienced a non-equilibrium transition from stable WC to equilibrium η phase, however, the period of the welding was very short and volatilization, if any, can occur mostly from the surface region. So the carbides were non-equilibrium phases. At the HAZ near WC-TiC-Ni/welds interface, carbon depletion phenomenon was observed and distributed widely. The average content (wt%) of SP3 (Fig. 6c) was 72.34W, 22.06Ni, and 3.68C, characterized with carbon depletion. Because the average contents of Ni and W in M_6C and $M_{12}C$ (M stands for Fe, Ni or Co) were nearly 23.2 and 75.30% respectively, η phase belonged to such compound carbides as M_6C and $M_{12}C$. In addition, W dissolution in Ni matrix (SP4 in Fig. 6d) was observed. The average contents (wt%) of SP4 were 9.34W, 64.27Ni, and 26.39Fe with carbon depletion. This meant the WC dissolution was going on accompanied by tungsten carbides decomposition into tungsten and carbon. The dissolution led to WC grains loss. Figure 6e, f showed the morphology of WC AGG. The average contents (wt%) of carbide (SP5, shown in Fig. 6e) and matrix (SP6, shown in Fig. 6f) were 94.32W, 5.68C and 16.11W, 18.40Fe, 56.88 Ni, 1.36C, 7.24Ti, which were characterized with carbon depletion.

3.4 Element Diffusion

Figure 7 showed the interfacial microstructure of specimens A1, A4, and A5 with EDS maps of as-welded joints. In the specimens A1, A4, and A5, the initial microstructures of cemented carbide (triangular, rectangular WC grains embedded in nickel binder) begin to dissolve to the fusion zone. WC dissolved in nickel with shape variation from triangular or rectangular to rounded WC grains accompanied by grain growth, which decreased the density of cemented carbide. Similar to the welded joint between cemented carbide and the steel, the microstructures of austenite and ferrite were the primary phases in HAZ and extended deeply into the steel. In the WC-TiC-Ni/welds interface, the HAZ consisted of austenite, γ-phase, η-phase, non-equilibrium carbide and WC. The fine dissolving WC grains were around the large WC grains. Vice versa, there were much more fine grains around the AGG WC grains due to the sacrifice of fine WC grains during WC dissolution and grain growth. The most notable features were undesirable voids and micro-cracking in the η phase reaction layer. The size of the voids and interfacial reaction layer were similar for all specimens. Such voids would act as origin of failure thus reducing mechanical robustness. These voids were due to the consequence of differential hot shrinkage induced by welding residual stress. WC AGG phenomenon occurred at interface in specimens A1 and A4 (Fig. 7b). However, in specimen A5 (Fig. 7c) there are less AGG WC grains. During robotic MIG welding, because the temperatures were too low to melt the WC grains, WC grains

Table 2 Composition distribution of WC behavior [wt% (at.%)]

Element	W M	Ni L	C K	Fe K	Ti K	Cr K
SP1	16.22 (4.95)	43.61 (41.68)	1.22 (5.69)	21.52(21.61)	9.45 (11.08)	5.36 (5.78)
SP2	40.50 (16.09)	44.71 (55.63)	1.87 (11.39)	12.91(16.89)	–	–
SP3	72.34 (32.91)	22.06 (31.44)	3.68 (25.61)	–	–	–
SP4	9.34 (3.14)	64.27 (67.66)	–	26.39(29.20)	–	–
SP5	94.32 (53.17)	–	5.68 (46.83)	–	–	–
SP6	16.11 (5.31)	56.88 (58.72)	1.36 (6.84)	18.40 (19.97)	7.24 (9.16)	–

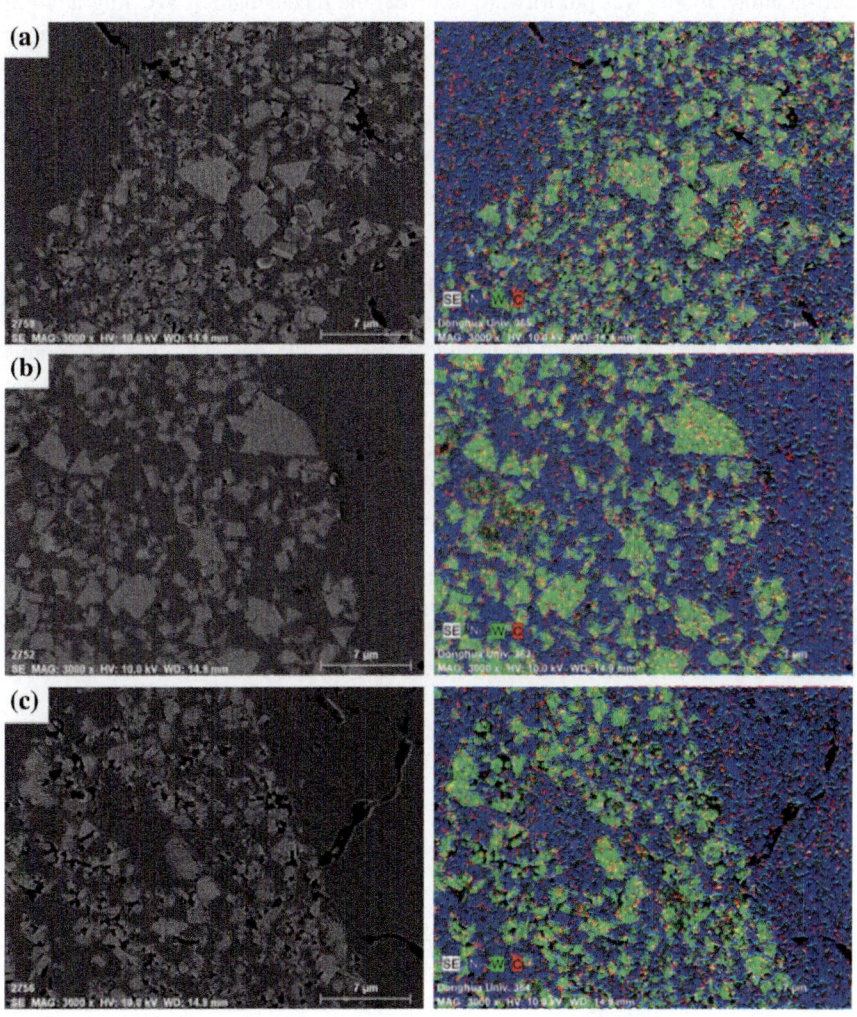

Fig. 7 SEM image and corresponding interfacial EDS compositional maps for area scanning micrograph of etched as-welded WC-TiC-Ni/304 stainless steel interface of specimens **a** A1; **b** A4; and **c** A5 (left: SEM image; right: EDS maps)

were not melted. However, under the effect of such forces as intrinsic stress, surface tension, buoyancy force, stirring behavior, and shock waves at the backside of weld pool, WC dissolution happened starting from the WC brims.

4 Discussion

4.1 *Effect of Heat Input on the WC Migration*

"Self-healing" model was put forward to reveal the mechanism of WC migration in welds. Figure 8a shows the schematic illustration of thermal process in an idealized vertical droplet, wire, and weld pool in WC-TiC-Ni. MIG welding melted and joined WC-TiC-Ni cemented carbide by heating them with an arc established between a continuously fed nickel wire and WC-TiC-Ni cemented carbide. The molten metals at the nickel wire tip were transferred to the weld pool by globular transfer mode across the arc gap under the effect of gravity. As shown in Fig. 8b, WC migration indicated a-type tungsten carbide (WC_a) migration in welds and b-type tungsten carbide (WC_b) migration in HAZ. WC_a grains had an uneven grain distribution; some WC_a grains were much larger than NGG WC grains, however, sizes of other WC_a grains were much smaller. There existed a dynamic equilibrium system consisted of fine WC_a grains, coarse WC_a grains, and liquid nickel matrix. On one side, fine WC grains dissolved to liquid nickel until it disappeared; on the other hand, WC nucleated at the surface of coarse WC grains at liquid/liquid or vapor/liquid interface at elevated temperature (shown as Figs. 5 and 6). WC_a featured with long-range moving with grain growth and interface I formation and interface II formation. The nickel appeared in HAZ came from both filler metal and base metal.

Heat input energy could be transferred to droplet and arc plasma, and then was transferred to workpiece. In the proposed model, energy loss q_l during robotic MIG welding: heat losses occur via conduction in the solid rod q_k and conduction in the workpiece q_w, radiation exchange with the environment q_r, convection to the shielding gas q_c, and other losses. So it can be expressed as,

$$q_l = q_r + q_c + q_k + q_w + \Delta q = (1 - \omega) \cdot Q \qquad (3)$$

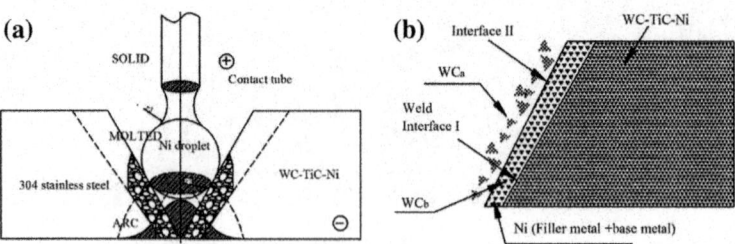

Fig. 8 Schematic illustration of **a** thermal process in an idealized vertical droplet, wire, and weld pool in WC-TiC-Ni; **b** Migration happened at heat affected zone

First, the governing thermal energy equation for energy transfer between nickel wire and droplet in cylindrical coordinates can be described as follows [31],

$$\frac{\partial \rho l}{\partial t} + \rho u \frac{\partial H}{\partial x} + \rho v \frac{\partial H}{\partial r} = \frac{\partial}{\partial x}\left(kr\frac{\partial T}{\partial x}\right) + \frac{1}{r}\frac{\partial}{\partial r}\left(kr\frac{\partial T}{\partial r}\right) + q_G''' \tag{4}$$

where ρ is the density, radial velocity v, axial velocity u, and the dependent variables are the enthalpy (H) and the specific internal thermal energy (l). r is radial distance; $r0$ is outside radius; $r1$ is the solid wire radius = 1.2 mm; k is thermo conductivity; x is axial distance; volumetric thermal energy generation rate q_G'''. Assume that steady motion of the solid wire, v is equal to zero, and u is equal to wire feed speed v_w. The heat transfer between nickel wire and droplet can be expressed as [31],

$$\frac{\partial \rho l}{\partial t} + \rho v_w u \frac{\partial H}{\partial x} = \frac{\partial}{\partial x}\left(kr\frac{\partial T}{\partial x}\right) + \frac{1}{r}\frac{\partial}{\partial r}\left(kr\frac{\partial T}{\partial r}\right) + q_G''' \tag{5}$$

Second, energy transfer between plasma arc (droplet) and weld pool can be expressed as follows,

$$q = \frac{\partial}{\partial t}(\rho H) + \nabla \cdot (\rho V H) \tag{6}$$

where, q is the total energy, in two or more dimensions we must use ∇. Continuum density, specific heat, thermal conductivity, solid mass fraction, liquid mass fraction, velocity, and enthalpy are defined as follows:

$$H = h_s f_s + h_l f_l \tag{7}$$

where f_s and f_l are the mass fracture of solid and liquid; h_s and hl are the enthalpy of solid and liquid. The direction of flow due to thermo capillary convection depends on whether the surface tension increases or decreases with temperature. In a sense, a region with higher stress σ pulls fluid from regions with lower stress σ. In addition, because the obvious difference of physical properties between WC-TiC-Ni cemented carbide and 304 stainless steel, the melting zone where warming liquid dropping toward will be determined by the above factors. In addition, the thermo capillary (Marangoni) has a significant effect on thermo-fluid-mechanics of weld pools. Surface tension generally depends on temperature, composition and electrical potential. The direction of flow due to thermo capillary convection depends on whether the surface tension increases or decreases with temperature.

In current research, the width of HAZ dH was used to estimate the WCb migration. This zone was adjacent to the welds and was not melted itself but was structurally changed because of the couple effect of thermal-mechanics. As shown in Fig. 9a, HAZ again was divided into two parts, the larger part neighboring the fusion zone containing nickel austenite, which was a quasi-equilibrium process in

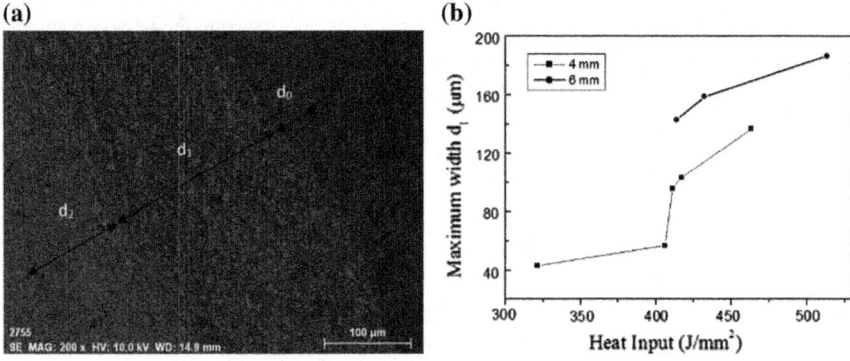

Fig. 9 a The definition of d_0-width of η phase layer; d_1-width of sparse layer; d_2-width of diffusion zone; **b** The relationship of d_1 and heat input H

liquid nickel during welding and dominated by dissolution. The width of this part was defined as d_1 along the vertical direction, which was related to the power density of heat source. The other part neighboring the WC-TiC-Ni cemented carbide and its width was defined as d_2 along the vertical direction. d_1 and d_2 were dependent variables and determined by the thermo-mechanics coupled. Figure 9b represented the results of relationship between d_0, d_1, d_2 and heat input during MIG welding of 4 and 6-mm thick specimens. Because the HAZ width was estimated from the process interaction time and the thermal diffusivity of the material [27], at levels above approximately 104 W/cm^2, the HAZ width became roughly constant. This was due to the fact that the HAZ grew during the heating stage at power densities that were below 104 W/cm^2, but at higher power densities it grew during the cooling cycle. Thus, at low power densities, the HAZ width was controlled by the interaction time, whereas at high power densities, the width was independent of the heat source interaction time.

WC$_a$ migration occurred in the region between HAZ and fusion zone. WC grains in weld pool reacted with the nickel from the melting filler rod, η phases were formed when nickel atoms substituted some tungsten atoms with carbon depletion. Too high carbon content led to the formation of graphite and too low carbon content led to the formation of an additional carbide, η phases (M_6C or $M_{12}C$) in quasi-equivalent state or transition carbide in non-equivalent [29, 30]. The difference in the possible loss of the carbon and tungsten due to volatilization must also be inconsiderable because the time period of the MIG welding was very short and volatilization, if any, could occur mostly from the surface region. During heating, among solid nickel wire and cemented carbide, molten droplet and weld pool, and arc plasma, droplet or arc plasma dominated the WC migration. Therefore, WC behavior included η layer formation; decomposition of WC → W + C, normal grain growth, and abnormal grain growth during WC$_a$ migration.

As described schematically in Fig. 10 for an idealized migrated WC, possible thermal processes that may interact with the forces on cemented carbide to cause WC$_b$ migration. For the ith ideal plane Γ_i in HAZ where WC migration occurred, its

Fig. 10 The stress applied on the idealized migrated WC and its failure mechanism

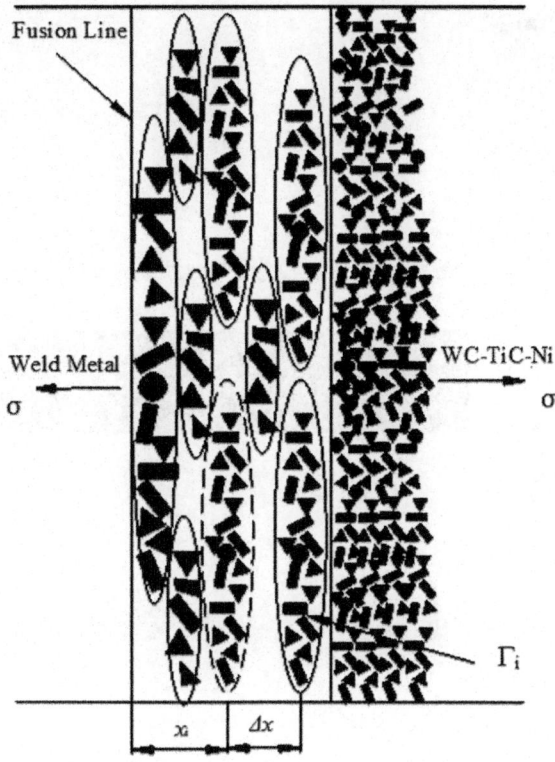

area is S; Δx_i and x are the distance of WC grains migration from its current position and from fusion line respectively. So the volume of WC in Γ_i is $S \cdot \Delta x_i \cdot f_{WC}$; σ is the total stress applied on the Γ_i containing migrated WC; σ_s is the yield strength. The energy for WC detachment from the original counterpart is $q_0(x)$, and the energy for WC migration per displacement is $q_m(x)$, so,

$$q_{WC} = S \cdot \Delta x \cdot f_{WC} \cdot q_m(x) + q_0(x) \tag{8}$$

Therefore,

$$\begin{cases} \frac{\partial}{\partial x}\left(kr\frac{\partial T}{\partial x}\right) + \frac{1}{r}\frac{\partial}{\partial r}\left(kr\frac{\partial T}{\partial r}\right) + q_G''' = \frac{\partial}{\partial t}(\rho H) + \nabla \cdot (\rho V H) = q_{WC} + Q \\ \sigma \geq \sigma_S \end{cases} \tag{9}$$

Criteria: assuming the number of WC grains is n in unit area, energy consumption for migration of per WC grain (WC_b) is $q_m(x)$ in unit distance along vertical direction. From the view of energy consumption, if energy provided by arc is greater or equal to the energy requirement for WC migration, that is, $q \geq q_{WC} + Q$, WC migration happens. In other words, when $\sigma \geq \sigma_s$, WC migration happens. When WC_b occurs at elevated temperature, liquid nickel can

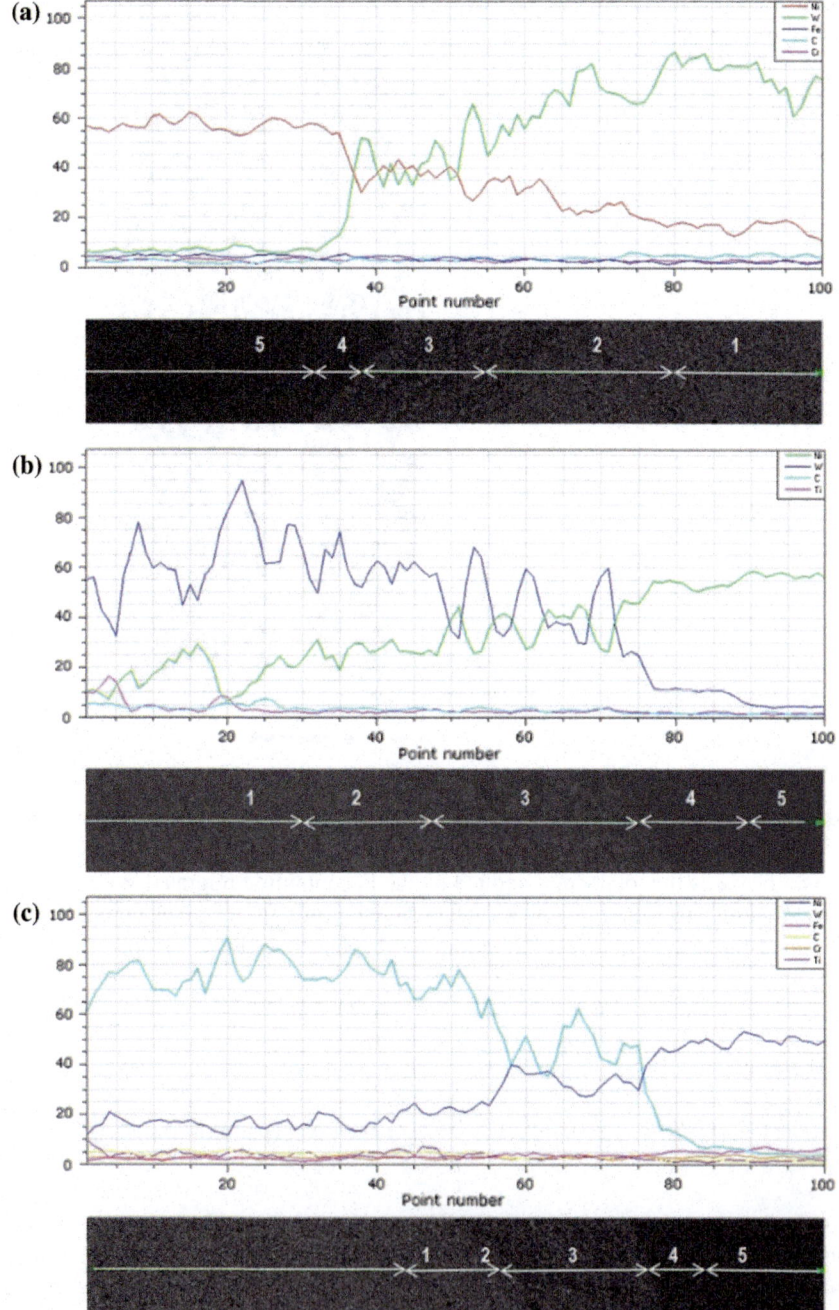

Fig. 11 Elements distribution at WC-TiC-Ni/Ni interface of specimens **a** A1; **b** A2, and **c** A5 (1. WC-TiC-Ni base metal; 2. HAZ; 3. Gradient layer; 4. η phase layer; 5. the fusion zone)

cure the cracks formation during welding. And the nearer to fusion line, the longer distance WC migrated.

According to elements distribution in the WC-TiC-Ni/welds interface, as shown in Fig. 11, approximate equilibriums were observed in region 3 (gradient layer) of specimens A1, A4, and A5, which agreed with previous investigation. However, in region 2 and region 4, there were obvious and consecutive transitions from the position with high concentration to that with the low concentration.

4.2 *Effect of Heat Intensity on the WC Migration*

If WC_a grains were characterized with electromagnetic-induced migration, the electromagnetically induced motion produced a rapid rate of transport of WC from HAZ into the weld pool, the formation of dispersive distribution of fine WC grains could be explained by the mechanism of dissolution-re-precipitation. Coarse WC grains and fine dispersive WC grains dissolved to liquid nickel which led to η formation. Because HAZ grew during the heating stage at power densities below $104 \ W/cm^2$ [31], WC_a grain migration was controlled by the interaction time, e.g. more WCa migration accompanied by normal grain growth and abnormal grain growth were observed in the fusion zones, which are mainly dominated by electromagnetic stirring effect. At higher power densities it grew during the cooling cycle. At high power densities, the width is independent of the heat source interaction time. The HAZ width grew during the cooling cycle as the heat of fusion was removed from the fusion zone and was proportional to the fusion zone width.

5 Conclusion

The amount and distance of WC_a migration in the fusion zone, for a given pure nickel welding wire, was determined by electromagnetically induced motion in the weld pool. This motion produced a rapid rate of transport of WC from HAZ into the fusion zone. WC_b migration indicated the stress levels to be above the yield stress in the fusion zone on-cooling during robotic MIT welding. WC behavior included the formation of η layer, decomposition of tungsten carbide, and grain growth during WC migration.

Tungsten was observed in the fusion zone near WC-Co side due to the dissolution and chemical decomposition of tungsten carbides. In the fusion zone, gradient layer was helpful for metallurgical bonding and it had the ability to cure the cracks.

Acknowledgements This work is financially supported by the National Natural Science Foundation of China (51475282, 51105240), the "Shu Guang" Project of Shanghai Municipal Education Commission and Shanghai Education Development Foundation (13SG54).

References

1. Exner HE (1979) Physical and chemical nature of cemented carbides. Int Met Rev 24: 149–173
2. Sui Y, Luo H, Lv Y et al (2016) Influence of brazing technology on the microstructure and properties of YG20C cemented carbide and 16Mn steel joints. Weld World 60(6):1269–1275
3. Guo Y, Wang Y, Gao B et al (2016) Rapid diffusion bonding of WC-Co cemented carbide to 40Cr steel with Ni interlayer: effect of surface roughness and interlayer thickness. Ceram Int 42(15):16729–16737
4. Klünsner T, Wurster S, Supancic P et al (2011) Effect of specimen size on the tensile strength of WC-Co hard metal. Acta Mater 59(10):4244–4252
5. Mirski Z, Granat K, Stano S (2016) Possibilities of laser-beam joining cemented carbides to steel. Weld Int 30(3):187–191
6. Lay S, Dennadieu P, Loubradou M (2010) Polarity of prismatic facets delimiting WC grains in WC-Co alloys. Micron 41(5):472–477
7. Ahn S, Kang S (2008) Dissolution phenomena in the Ti $(C_{0.7}N_{0.3})$-WC-Ni system. Int J Refract Metal Hard Mater 26(4):340–345
8. Li LQ, Liu DJ, Chen YB et al (2009) Electron microscopy study of reaction layers between single-crystal WC particle and Ti-6Al-4V after laser melt injection. Acta Mater 57(12):3606–3614
9. Barbatti C, Garcia J, Liedl G et al (2007) Joining of cemented carbides to steel by laser beam welding. Materialwiss Werkstofftech 38(11):907–914
10. Lavergne O, Robaut F, Hodaj F et al (2002) Mechanism of solid-state dissolution of WC in Co-based solutions. Acta Mater 50(7):1683–1692
11. Cheniti B, Miroud D, Badji R et al (2017) Effect of brazing current on microstructure and mechanical behavior of WC-Co/AISI 1020 steel TIG brazed joint. Int J Refract Metal Hard Mater 64:210–218
12. Su JZ, Guo LJ, Bao NZ et al (2011) Nanostructured $WO_3/BiVO_4$ heterojunction films for efficient photo electro-chemical water splitting. Nano Lett 11(5):1928–1933
13. Zou J, Sun SK, Zhang GJ et al (2011) Chemical reactions, anisotropic grain growth and sintering mechanisms of self-reinforced ZrB_2-SiC doped with WC. J Am Ceram Soc 94(5):1575–1583
14. Guo J, Fang ZZ, Fan P et al (2011) Kinetics of the formation of metal binder gradient in WC-Co by carbon diffusion induced liquid migration. Acta Mater 59(11):4719–4731
15. Zhang XZ, Liu GW, Tao JN et al (2017) Vacuum brazing of WC-8Co cemented carbides to carbon steel using pure Cu and Ag-28Cu as filler metal. J Mater Eng Perform 26(2):488–494
16. Niu L, Hojamberdiev M, Xu YH (2010) Preparation of in situ-formed WC/Fe composite on gray cast iron substrate by a centrifugal casting process. J Mater Process Tech 210(14): 1986–1990
17. Konyashin I, Hlawatschek S, Ries B et al (2009) On the mechanism of WC coarsening in WC–Co hard metals with various carbon contents. Int J Refract Metal Hard Mater 27(2): 234–243
18. Upadhyaya GS (1998) Cemented tungsten carbides: production, properties, and testing. Noyes Publications, Westwood
19. Fernandes CM, Senos AMR, Vieira MT (2007) Control of η carbide formation in tungsten carbide powders sputter-coated with (Fe/Ni/Cr). Int J Refract Metal Hard Mater 25(4): 310–317
20. Mercado WB, Cuevas J, Castro IY et al (2007) Synthesis and characterization of Fe_6W_6C by mechanical alloying. Hyperfine Interact 175(1–3):49–54
21. Suryanarayana C (2001) Mechanical alloying and milling. Prog Mater Sci 46(1–2):1–184
22. Wang Y, Zhu N, Wang H et al (2017) Interdiffusion and diffusion mobility for fcc Ni-Co-Al alloys. Metall Mater Trans A 48(3):943–947

23. Weidow J, Johansson S, Andrén HO et al (2011) Transition metal solubilities in WC in cemented carbide materials. J Am Ceram Soc 94(2):605–610
24. Liu S, Huang Z, Liu G et al (2006) Preparing nano-crystalline rare earth doped WC/Co powder by high energy ball milling. Int J Refract Metal Hard Mater 24(6):461–464
25. Rios O, Cupid DM, Seifert HJ et al (2009) Characterization of the invariant reaction involving the L, η, γ and σ phases in the Ti–Al–Nb system. Acta Mater 57(20):6243–6250
26. Kou S (2002) Welding metallurgy, 2nd edn. Wiley, Hoboken, pp 205–418
27. Kim CS, Massa TR, Rohrer GS (2008) Interface character distributions in WC–Co composites. J Am Ceram Soc 91(3):996–1001
28. ASM (2004) ASM handbook volume 9: metallography and microstructures. ASM International, Materials Park, OH, pp 125–249
29. Mannesson K, Jeppsson J, Borgenstam A et al (2011) Carbide grain growth in cemented carbides. Acta Mater 59(5):912–1923
30. Yang DY, Yoon DY, Kang SJL (2011) Suppression of abnormal grain growth in WC–Co via two-step liquid phase sintering. J Am Ceram Soc 94(4):1019–1024
31. Kim YS, Mceligot DM, Eagar TW (1991) Analyses of electrode heat transfer in gas metal arc welding. Weld J 70(1):20-s–31-s

Part III
Short Papers and Technical Notes

Study on Properties of LBW Joint of AISI 304 Pipes Used in Nuclear Power Plant

Jingyao Chen, Xiao Wei, Jijin Xu, Yi Duan, Junmei Chen, Chun Yu and Hao Lu

Abstract In this work, a girth weld of AISI 304 stainless steel pipes with a dimension of Φ89 mm × 250 mm × 14.5 mm was produced by laser beam welding. The microstructure was observed by optical microscopy. The microhardness test was carried out to study the mechanical properties of the welded joint. A slow strain rate test was performed to investigate the susceptibility to stress corrosion cracking (SCC) of the welded joint in the simulated pressurized water reactor environment. The results show that the values of microhardness in the weld and HAZ are close to those in the parent material. According to ultimate tensile strength-loss rate, elongation-loss rate and fractograph analysis, the welded joint has low susceptibility to SCC in the simulated pressurized water reactor environment.

Keywords Laser beam welding · Slow strain rate test · SCC susceptibility Microhardness · AISI 304 stainless steel

1 Introduction

AISI 304 austenitic stainless steel has been widely used in the fabrication industries including nuclear power industry because of its excellent weldability, good mechanical characteristics and corrosion resistance properties [1]. It is a significant structural material of the pressurized water reactor, which is commonly fabricated by welding techniques.

J. Chen · X. Wei · J. Xu (✉) · Y. Duan · J. Chen (✉) · C. Yu · H. Lu
Key Lab of Shanghai Laser Manufacturing and Materials Modification,
School of Materials Science and Engineering, Shanghai Jiao Tong University,
200240 Shanghai, China
e-mail: xujijin_1979@sjtu.edu.cn

J. Chen
e-mail: chenjm@sjtu.edu.cn

© Springer Nature Singapore Pte Ltd. 2018
S. Chen et al. (eds.), *Transactions on Intelligent Welding Manufacturing*,
Transactions on Intelligent Welding Manufacturing,
https://doi.org/10.1007/978-981-10-8330-3_10

Laser beam welding (LBW), as a new welding technology, is widely developed and applied in many manufacturing industries due to its excellent properties including low heat input, small heat affected zone, high producing efficiency and deep penetration. However, welding leads to low mechanical properties owing to the metallurgical changes during the welding process [2, 3], which will affect the performances of welded structures in services, such as the occurrence of stress corrosion cracking (SCC).

SCC can lead to accidents during the long-term services, which has become a serious problem in the nuclear power industry [4]. The SCC initiates and grows near the welding zone because of high tensile residual stress by welding same as the other contributing factors of material and environment [5]. It is caused and accelerated by the presence of corrosive environments, residual stress, and material sensitization effects [6]. Plenty of researches [7, 8] have been reported to study the SCC behavior of stainless steel in corrosive environment.

Slow strain rate test (SSRT) is a widely used method on stress corrosion cracking researches as the basic experimental technique to promote the incidences of cracking and assess the ranking of susceptibility of different alloys in several corrosive environments. However, few research which focused on the SCC behavior of thick pipeline joints induced by LBW was reported.

In this work, a girth weld of AISI 304 stainless steel thick pipes was produced by LBW. The microstructure of joint was observed by optical microscopy. The microhardness test was carried out to study the mechanical properties of the joint. A slow strain rate test was performed to investigate the susceptibility to SCC of the welded joint in the simulated pressurized water reactor environment.

2 Experimental Procedures

2.1 Materials and Welding Process

The dimension of AISI 304 pipes is $\Phi89$ mm $\times$ 250 mm $\times$ 14.5 mm. The composition of parent material was shown in Table 1. Table 2 shows the welding parameters. During welding, temperature histories were recorded with thermocouples located on the inner surface of pipe, 4 mm to the weld center. After Welding, the axial shrinkage was measured based on fixed point method.

After polished by emery paper, the specimen was electrolytic etched by oxalic acid, which are composed of 10 g oxalic and 90 g distilled water. Afterwards, the microstructures of the joint were captured by an optical microscope.

Table 1 Chemical compositions of parent metal and filler (wt%)

Materials	C	Mn	Cr	S	Ni	Si	P	Fe
304	0.06	1.89	18.67	0.014	8.53	0.42	0.032	other

Table 2 The welding parameters

Methods	Groove type	Weld pass	Power	Rate	Shield gas
LBW	I-type	Single-pass	11 kW	8 mm/s	N_2

2.2 Microhardness Test

The microhardness of the joint was measured by Zwick automatic hardness tester in the longitudinal direction. Each test point was evaluated with a testing load of 300 g and was measured with an interval of 0.5 mm.

2.3 Slow Strain Rate Test

Rod like samples were sliced from welded pipelines along the longitudinal direction and ensure the weld metal at the center of specimens. The gauge length is 25.4 mm and the gauge diameter is 4 mm.

The SCC susceptibility of the joints was studied by SSRT with a strain rate of 4×10^{-7}/s in a simulated primary water environment at 350 °C and 11.5 MPa without dissolved oxygen. The simulated primary water contained 1200 ppm B as boric acid and 2.2 ppm Li as lithium hydroxide. The fractography of specimens after fracture was characterized by JSM-7800F scanning electron microscope (SEM).

3 Results and Discussion

3.1 Microstructure

Macroscopic cross-section of the specimen and the microstructure of different regions, including weld metal (WM), fusion zone (FZ), heat affected zone (HAZ) and parent metal (PM) in the joint, are shown in Fig. 1. The joint shows no cracks and porosities from Fig. 1a. It shows narrow weld width and small fusion area due to high welding speed and low heat input during welding process. The fusion area is 57.61 mm². The axial shrinkage of the joint can be calculated by the empirical formula as follows:

$$\Delta y = 0.2 \frac{A_H}{\delta} \tag{1}$$

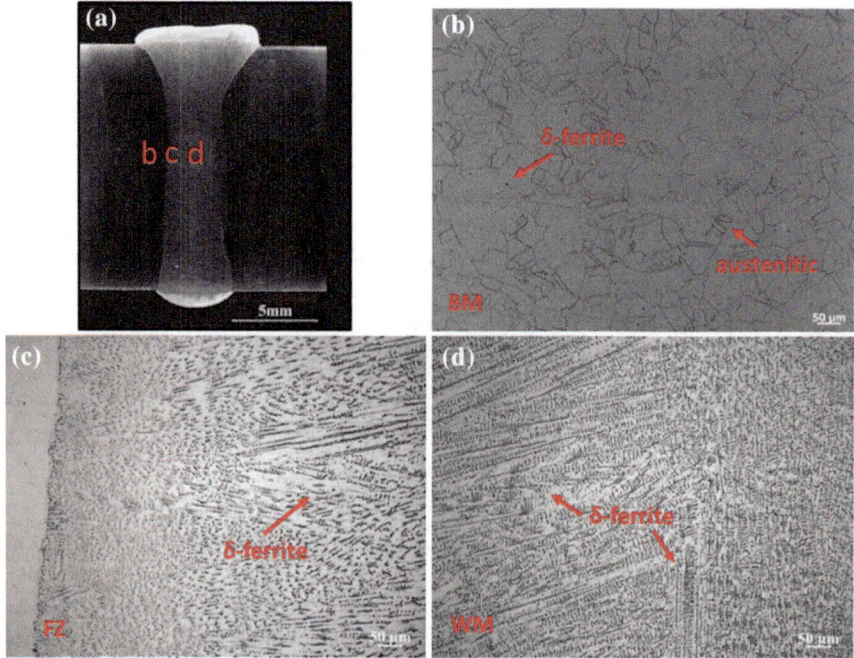

Fig. 1 Microstructure of **a** Overview **b** BM **c** FZ **d** WM

Δy is the axial shrinkage; A_H is the fusion area and δ is the thickness of the pipes. The calculated axial shrinkage of the joint is 0.79 mm, which is close to the measured value, 0.56 mm [9].

According to Fig. 1b–d, the parent metal of AISI 304 stainless steel is composed of equiaxed austenitic grains and a small volume fraction of δ-ferrite. While the weld metal of the joint is composed of austenitic matrixes and plenty of δ-ferrite dendrites, which means that lots of δ-ferrite produced during welding process remained. The solidification process in welding is a non-equilibrium and fast process and the temperature in fusion zone is higher than the temperature of phase balance between δ-Fe and γ-Fe phase so that a lot of δ-ferrite occurred in joint [10]. The dendrites grow from weld metal to fusion line because the direction of grain growth is the opposite to the direction of heat dissipation.

3.2 Microhardness

The microhardness of the joint was measured along the transverse direction parallel to the surface. The results were shown in Fig. 2. Temperature histories during welding are presented in Fig. 3.

Fig. 2 Distribution of hardness in longitudinal direction

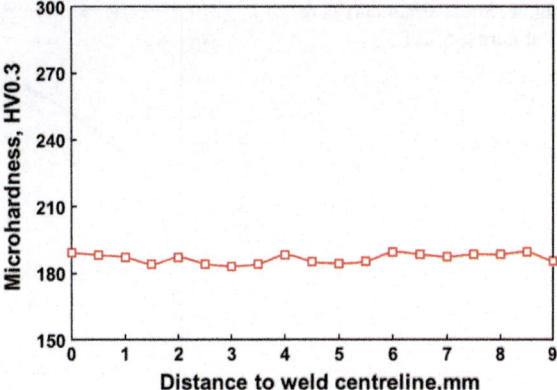

Fig. 3 Welding thermal cycle of the measured point of the joint

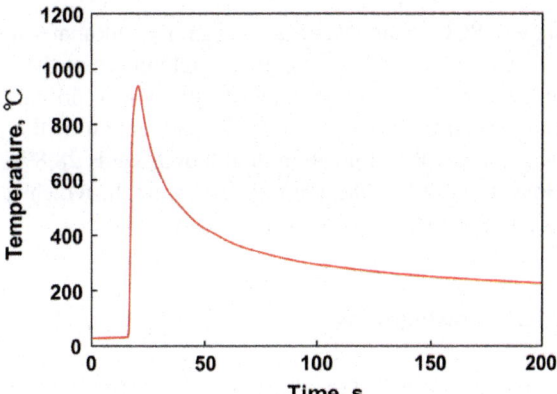

From Fig. 2, it is observed that microhardness changes little along the transverse direction and the values of weld metal and HAZ were close to the hardness of parent metal. The $t_{8/5}$ of the measured point is about 16 s according to Fig. 3. The cooling time of the joint from 800 to 500 °C is larger than usual because of the thickness of the pipes and it reveals higher heat input during welding process which results in similar microhardness in weld metal and parent metal.

3.3 SSRT

3.3.1 Stress-Strain Curves

The stress-strain curves of the SSRT are presented in Fig. 4. The results show that the yield strength and ultimate tensile strength in water environment are lower than the tensile properties in air condition. The tensile properties depression of the joint

Fig. 4 Stress-strain curves of joint during SSRT

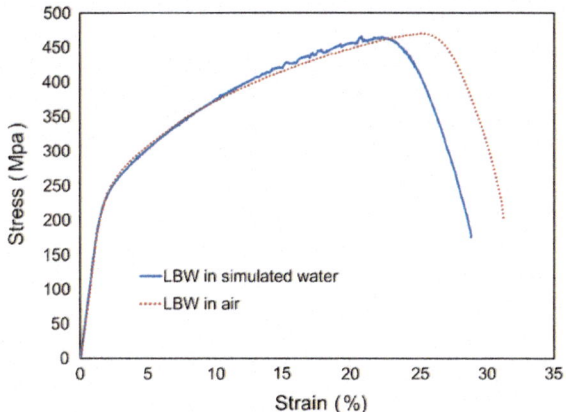

shows SCC characteristics. Generally, ultimate tensile strength-loss rate and elongation-loss rate are significant references of the SCC susceptibility. According to the curves, the ultimate tensile strength in primary water is 466 MPa while the value in air is 470 MPa. Also, the elongation of the joint in water environment is 31.2% while the elongation in air condition is 28.8%. The ultimate tensile-loss and elongation-loss of the joint are much small, which reveals that the SCC susceptibility of the laser welded joint is tiny.

3.3.2 Fractography

The fractured surfaces of the tested specimens were analyzed using JSM-7800F SEM. The fractographs of the joints after SSRT in air and in simulated primary water were shown in Fig. 5.

Figure 5a, b show the fractographs under different magnification of the joints after SSRT in air condition. The fractographs of the joint consist of much dimples both in the center and at the rim of surface, which shows definite ductile fracture characteristic.

As shown in Fig. 5c, d, the fractographs of the joint tested in the simulated primary water environment present two kinds of features. The center of fracture surface is covered by deformation dimples and reveals the feature of ductile fracture, while the cleavage platform observed at the rim of the fracture surface presents the feature of brittle fracture. The proportion of cleavage fracture regions in the surface reflected the SCC susceptibility of joints. From the fractographs, the proportion of brittle fracture regions of the joint is much small which means its high resistance to SCC.

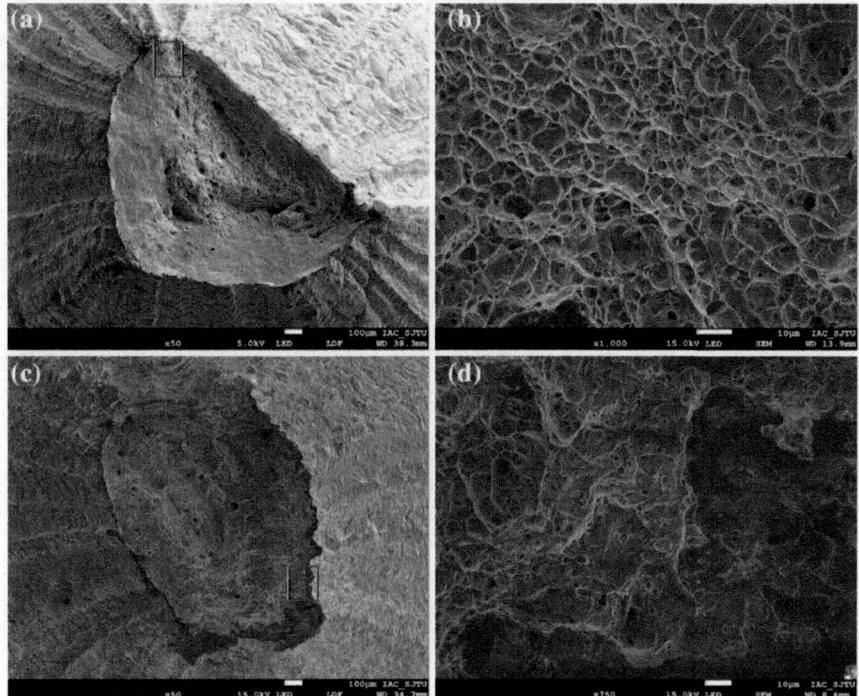

Fig. 5 SEM fractographs of joint in air **a** center area at 50X; **b** edge area at 1000X and in simulated primary water; **c** center area at 50X; **d** edge area at 1000X

4 Conclusion

The microstructure reveals that weld metal is composed of austenitic matrix and δ-ferrite dendrites due to non-equilibrium cooling process. The micro-hardness of the weld metal and HAZ are close to the value of the parent material.

The yield strength and ultimate tensile strength of the joint in simulated primary water are lower than that in air condition. The joint shows good resistance to SCC in simulated primary water environment.

References

1. Subodh K, Shahi AS (2011) Effect of heat input on the microstructure and mechanical properties of gas tungsten arc welded AISI 304 stainless steel joints. Mater Des 32:3617–3623
2. Lee HT, Jeng SL (2001) Characteristics of dissimilar welding of alloy 690 to 304L stainless steel. Sci Technol Weld Joining 6:225–234

3. Jamshidi AH, Farzadi A, Serajzadeh S et al (2009) Theoretical and experimental study of microstructures and weld pool geometry during GTAW of 304 stainless steel. Int J Adv Manufact Technol 42:1043–1051

4. Wolfgang R, Mary PN, George SG (2011) Civilian nuclear accidents. Encycl Inorg Bioinorg Chem 15:1–14

5. Jinya K (2007) Crack growth analysis of SCC under various residual stress distributions near the piping butt-welding. In: ASME Pressure vessels and piping division conference, vol 1, pp 22–26

6. Scott PM (2007) An overview of materials degradation by stress corrosion in PWRs. Eur Fed Corros Publ 1:3–24

7. Takumi T, Katsuhiko F, Koji A (2005) Microstructural characterization of SCC crack tip and oxide film for SUS 316 stainless steel in simulated PWR primary water at 320 °C. J Nucl Sci Technol 42:225–232

8. MacDonald DD, Urquidi-MacDonald M (1991) A coupled environment model for stress corrosion cracking in sensitized type 304 stainless steel in LWR environments. Corros Sci 32:51–81

9. Xu JJ, Chen JY, Duan Y et al (2017) Comparison of residual stress induced by TIG and LBW in girth weld of AISI 304 stainless steel pipes. J Mater Process Technol 248:178–184

10. Muthupandi V, Bala SP, Seshadri SK et al (2003) Effect of weld metal chemistry and heat input on the structure and properties of duplex stainless-steel welds. Mater Sci Eng A 358:9–16

Information for Authors

Aims and Scopes

Transactions on Intelligent Welding Manufacturing (TIWM) is authorized by Springer for periodical publication of research papers and monograph on intelligentized welding manufacturing (IWM).

The TIWM is a multidisciplinary and interdisciplinary publication series focusing on the development of intelligent modelling, controlling, monitoring, and evaluating and optimizing the welding manufacturing processes related to the following scopes:Scientific theory of intelligentized welding manufacturing

- Scientific theory of intelligentized welding manufacturing
- Planning and optimizing of welding techniques
- Virtual & digital welding /additive manufacturing
- Sensing technologies for welding process
- Intelligent control of welding processes and quality
- Knowledge modeling of welding process
- Intelligentized robotic welding technologies
- Intelligentized, digitalized welding equipment
- Telecontrol and network welding technologies
- Intelligentized welding technology applications
- Intelligentized welding workshop implementation
- Other related intelligent manufacturing topics

Submission

Manuscripts must be submitted electronically in WORD version on online submission system: https://ocs.springer.com/ocs/en/home/TIWM2017. Further assistance can be obtained by emailing Editorial Office of TIWM, Dr. Yan ZHANG: zhangyan521@sjtu.edu.cn, or one of the Editors-in-Chief of TIWM.

Style of Manuscripts

The TIWM includes two types of contributions in scopes aforementioned, the periodical proceedings of research papers and research monographs. Research papers include four types of contributions: Invited Feature Articles, Regular Research Papers, Short Papers and Technical Notes. It is better to limit the

© Springer Nature Singapore Pte Ltd. 2018
S. Chen et al. (eds.), *Transactions on Intelligent Welding Manufacturing*,
Transactions on Intelligent Welding Manufacturing,
https://doi.org/10.1007/978-981-10-8330-3

full-length of Invited Feature Articles in 20 pages; Regular Research Papers in 12 pages; and Short Papers and Technical Notes both in 6 pages. The cover page should contain: Paper title, Authors name, Affiliation, Address, Telephone number, Email address of the corresponding author, Abstract (100–200 words), Keywords (3–6 words) and the suggested technical area.

Format of Manuscripts

The manuscripts must be well written in English and should be electronically prepared preferably from the template "splnproc1110.dotm" which can be downloaded from the website: http://rwlab.sjtu.edu.cn/tiwm/index.html. The manuscript including texts, figures, tables, references, and appendixes (if any) must be submitted as a single WORD file.

Originality and Copyright

The manuscripts should be original, and must not have been submitted simultaneously to any other journals. Authors are responsible for obtaining permission to use drawings, photographs, tables, and other previously published materials. It is the policy of Springer and TIWM to own the copyright of all contributions it publishes and to permit and facilitate appropriate reuses of such published materials by others. To comply with the related copyright law, authors are required to sign a Copyright Transfer Form before publication. This form is supplied to the authors by the editor after papers have been accepted for publication and grants authors and their employers the full rights to reuse of their own works for noncommercial purposes such as classroom teaching etc.

Author Index

© Springer Nature Singapore Pte Ltd. 2018
S. Chen et al. (eds.), *Transactions on Intelligent Welding Manufacturing*,
Transactions on Intelligent Welding Manufacturing,
https://doi.org/10.1007/978-981-10-8330-3